Deep Learning in Time Series Analysis

Arash Gharehbaghi

Researcher, School of Information Technology
Halmstad University, Halmstad, Sweden

CRC Press
Taylor & Francis Group
Boca Raton London New York

CRC Press is an imprint of the
Taylor & Francis Group, an **informa** business

A SCIENCE PUBLISHERS BOOK

Cover illustration courtesy of Reza Gharehbaghi

First edition published 2023
by CRC Press
6000 Broken Sound Parkway NW, Suite 300, Boca Raton, FL 33487-2742

and by CRC Press
4 Park Square, Milton Park, Abingdon, Oxon, OX14 4RN

Library of Congress Cataloging-in-Publication Data (applied for)

ISBN: 978-0-367-32178-9 (hbk)
ISBN: 978-1-032-41886-5 (pbk)
ISBN: 978-0-429-32125-2 (ebk)

DOI: 10.1201/9780429321252

Typeset in Times New Roman
by Radiant Productions

To Shabnam, Anita and Parsa.

Foreword

I am delighted to introduce the first book on deep learning for time series analysis in which analysis of cyclic time series is profoundly addressed along with the theories. This idea was developed within a structure of a hybrid model where the experimental results showed its outperformance against the baselines of neural network-based methods. It was later improved by incorporating deep learning structures of a time growing neural network, the network which was previously introduced by us as a strong alternative to multilayer perceptron and time-delayed neural networks, into a multi-scale learning structure. The idea of cyclic learning is applicable to many natural learning where the phenomena exhibit cyclic behaviours. Physiological characteristics of the human body emanate cyclic activities in many cases such as cardiac and respiratory activities. The idea of cyclic learning has received interest from researchers from various domains of engineering and science.

Realistic validation of machine learning methods is a crucial task. A realistic validation method must provide sufficient outcomes to project capability of a machine learning method in terms of its risks in reproducibility of the results in conjunction with the improvement of the results when the machine learning method is being trained by a richer dataset. These validation capabilities are considered in the A-Test method. As a validation method, A-Test has received recognition from different engineering domains. These methods are likely to become strong machine learning methods, especially for applications with a small size of the learning data.

Preface

Learning has been regarded as an important element of development by most of the scientific pursuits including computer sciences in which deep machine learning has recently sounded as an emerging context. Application of deep machine learning methods has been well-received by the researchers and engineers since the last half decade when time series analysis was increasingly regarded as an important topic within different contexts such as biomedical engineering. Although development of strong tools for the implementation of deep learning methods created a breakthrough in computer science and engineering, a shift towards abstract understanding within this context is clearly seen, especially in the younger developers. This can put a negative impression on the general beliefs of deep learning which will be in turn considered as downside of this progress. Nowadays, various deep learning methods are enormously developed and published in the highly reputable references, however, a very low percentage of them entail sufficient quality to make a real impact on the underlying community. One reason can be the lack of sufficient insight into the theoretical foundation as well as into the implementation knowledge. This motivated the author to prepare a textbook on deep learning methods, sophisticated for time series analysis, to bring up fundamentals of the context along with the algorithms for the implementation.

Book Focus

This book focuses on the deep notions of the learning process in general, and deep learning in particular, with more orientation towards essentials of the traditional methods and the modern ones for time series analysis. Although image processing is known as an important topic of deep learning, the authors concluded to exclude this topic from the book and assign it to a separate publication as future work. The rationale behind this conclusion was mainly to avoid extra diversity and losing central attention. The book begins with a smooth transition from the fundamental definitions and speculations toward method formulation. Contents of the book were pedagogically organized in a way to foster and consolidate the essentials of time series analysis. This manner of representation is set to broaden the scope of the readers from the scientific to the engineering aspects. The book also considered to bring up a number of the practical examples of the deep learning methods for time series analysis, with the rise of biomedical engineering and medical informatics applications. Meanwhile, the book represents the deep learning methods mostly in a mathematical manner to help the researchers and the developers in mathematically formulating their own methods. It is evident that mathematical representation of a new method provides better readability compared to the descriptive representation. It is seen that new students show more tendency to learn concepts of deep learning using block diagrams and descriptive methods. Indulging in this learning manner can mislead them from the basic abilities in mathematical representation that can act as a degenerative factor for the learning of deep learning methods. Furthermore, a consistent graphical representation is not seen in many cases.

A number of the new ideas in artificial intelligence are also presented in this book. The A-Test validation method is introduced and compared to the other traditional ones. The readers can easily find out elaborations of this method in providing a more realistic validation as compared to the other two alternatives. In terms of the learning models, the idea of cyclic time series and cyclic learning are the other two new concepts addressed by this book and some of the learning methods such as time growing neural network is also introduced for learning cyclic time series.

Generative models for time series analysis were not addressed by this book. These models fit well into the prediction category which is considered as part of the future work.

Book Readership

This book, as a textbook, has been written in a fashion to establish fundamentals of time series analysis and deep learning methods for the readers. Problem formulation, as well as the methodological representations, have been rendered in a way to address the notional contents with a special focus on the scientific manners, so all the students, scientists, engineers and developers who are interested to learn deep learning methods for time series analysis or building up their own heuristic methods, can find this book interesting to read. The students in engineering, particularly those in artificial intelligence, are rather encouraged to read this book as a textbook.

Contributions

The book title as well as the arrangement of the chapters and the chapter titles were prepared by Arash Gharehbaghi and Ning Xiong. Contents of the first 10 chapters have been completely written by Arash Gharehbaghi. It incorporates all the writings, graphical representations, and tables. Chapter 11 was prepared by Ning Xiong and Johan Holmberg. Chapter 12 was prepared by Elaheh Partovi and Ning Xiong.

Arash Gharehbaghi

Contents

Contributors

Arash Gharehbaghi

School of Information Technology, Halmstad University, Halmstad, Sweden.

Ning Xiong

School of Innovation, Design and Technology, Mälardalen University, Västerås, Sweden.

Johan Holmberg

School of Innovation, Design and Technology, Mälardalen University, Västerås, Sweden.

Elaheh Partovi

Department of Electrical Engineering, Amir Kabir University (Tehran, Polytechnique), Tehran, Iran.

Part I

Fundamentals of Learning

1

Introduction to Learning

All the components of nature have been set in an everlasting pursue towards evolution through a growing network of mutual effects. Evolution is defined as an absolute point of dimensionless in time when the fully interconnected network of effects is moving in a completely predictable manner. The pursue towards evolution incrementally occurs in the form of an adaptation, not only for brainless materials, but also for any kind of intelligent creatures, such that adaptability to the underlying environment is defined as one of the indicative parameters of intelligence. Adaptation of a component can be expressed by the inherent temporal changes in the behaviour, and ultimately in the structure of the component to exhibit further similarities with the surrounding environment. This important definition will be seen in the section about the learning theory in Chapter 2.

At this point, one should consider time as an important element of adaptation. In fact, adaptation is a dynamic process, through which the adapting system adjusts, itself to the surrounding environment which is, in turn, another dynamic element. In a dynamic environment, an adaptation of a component is seen as a reaction of the component to the environmental variations towards the same objective, exhibiting further similarities with the surrounding environment. However, interpretation of this adaptation varies when it comes to computer science in which decision making is regarded as an important feature of intelligence. Although this point of view might change when it comes to the psychological perspectives, in a broader sense, decision making is a way of adaptation. These two scopes converge to a single point if we consider the process leading to decision making, which is based on learning similarity along with the differences over various groups of data, which is by itself linked to adaptation strongly. Putting such the absolute definitions (which may not be observed within the lifetime of components) into consideration, can be intuitive for real-world scenarios both for the problem formulation, and for the solution methodologies, as like as the development of the deductive sciences in which absolute definitions provide a theoretical foundation of the applicative contents. In analogy, a similar development has happened in mathematics, when the analytical mathematics provided fundamentals of the numerical mathematics to respond to the complicated real-world questions.

Numerical mathematics has been well-embraced since the development of digital computers [127]. It is obvious that many machine learning methods are based on the methods initiated by the scientists from the numerical mathematics domain, without which one could barely imagine such a rapid progress in artificial intelligence. It can be concluded that the natural development of deductive science is typically initiated by the analytic foundation of the theories, and extends towards application, unlike many branches of natural sciences which flow from observation to the theories. It is almost customary in chemistry that a phenomenon is observed first, and then the scientists bring up theories to explain the phenomenon. In other words, the incentive begins from an observation of the theory. In mathematics, the journey sometimes oppositely commenced from a theory to the application. This attitude has been tried to be followed in this book. The principals and concepts are described before methodologies to link the readers to the deep notion of the methods. The authors believe that it is essential for the readers to deeply understand the notions, logic, and reasoning, hidden behind each presenting method.

Nowadays, numerous and varied methods along with the pertinent open source codes for the implementation, have exploded within the community of artificial intelligence, thanks to the new advances in computer engineering [76]. Certainly, a broad range of options are being opened to researchers, which is naturally favourable, however, selecting the most appropriate one among the innumerable options, is not an easy task if one suffices to the practical aspects only. Furthermore, there are always hyper-parameters (sometimes called design parameters), associated with each method, and therefore providing an optimal solution for a research question is almost unpractical without a deep understanding of the theoretical foundation of the method. Many experienced scientists believe scientific studies were averagely deeper in analytic methods before the popularization of the open source implementation of the methods, when the scientists had to develop their codes! Such tough judgement is of course controversial and out of the belief of the authors, but worth considering to conclude; a deep understanding of the deep learning methods is essential prior to any kind of implementation.

Deep learning is indeed a sophistication of the above-mentioned adaptation. Learning theory would be described in more detail in the next chapter, nevertheless, it is worth addressing one important link between learning and adaptation in this instance. Learning is mostly concomitant with decision making, for better compliance with a minimal number of parameters. In contrary, adaptation is mostly a continuous process of system parameters to show better similarity to a specific reference. This point will be expanded in more detail in the section about learning theory, but it is important to note that both learning and adaptation processes depend on the environmental dynamics or the input data. To provide a consistent presumptions for the rest of the chapters, some of the definitions are described in the following sequels. These definitions will be presented descriptively, starting from the point of signal, data, and eventually time series, definition, and landing for analysis. Fundamentals of noise and the existing models for the analysis will be addressed as well. It is important to establish the definitions in a clear way at the introduction, as most of the theories and methodologies are built upon these bases. This chapter will terminate with a brief view of the book organisation.

1.1 Artificial Intelligence

Adaptation of a natural element is affected by a superposition of the surrounding elements through a network of interactions. Many scientists believe in a high level of intelligence holistically governing natural movements. However, a partial insight into the tiny particles shows that intelligence in an element begins when a level of decision making occurs in that element, is not generally true [4]. An approach towards simulating such an intelligence by using mathematical tools, is known as "artificial intelligence". The conventional approach for a deeper understanding of intelligence was firstly inspired by the human brain and concentrated on the learning process only [69]. The presented model became a frontier against statistical methods which were already a popular common, as a more powerful alternative for the learning, however, researchers found out that the two alternatives were intrinsically similar but algorithmically different in terms of the calculation [15]. As a consequence, the gap between the statistical way of learning and artificial intelligence-based fashion became narrower, such that the two alternatives became well merged [115]. In contrast with the traditional view of artificial intelligence which was predominantly about the learning process, modern perspective has broadened the scope of artificial intelligence to an expansive context including three main topics shown in Figure 1.1.

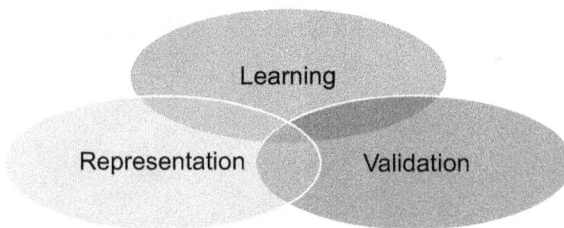

FIGURE 1.1: The three main topics of artificial intelligence.

In this perspective, learning is limited to only one topic of the larger context of artificial intelligence against the two other topics: representation and validation. If the learning is methodologically based on extracting information from the data through procedural machine-based routines, the learning is named "machine learning". Representation addresses methods for demonstration, dimension reduction or quantification of information. Validation is concerned with evaluating the performance of the learning process, which is in turn affected by the representation. Nevertheless, as we will see in the upcoming chapters, in the modern learning perspectives, validation sometimes influences the learning, or even the representation processes.

1.2   Data and Signal Definition

Data is defined as, AN ASSOCIATION TO AN EXISTENCE! This definition contains two keywords: association and existence. Association can be attribution, defined according to the findings, e.g., facial colour of a patient. It is often collected by the measurement, numerical or symbolical label, or obtained through mathematical or statistical mapping. Existence can be a phenomenon, an object, a living object, a signal, or any kind of image. For example, a person's weight is a measuring data, while the facial colour of a newborn baby is a symbolic data collected from the individuals. Weight and colour are the association and the individual from which the data is collected is the existence. In science and engineering, the signal is a sequential registration or/and representation of a phenomenon in time. The phenomenon is often represented in time-value axes. As an example, temporal variation of the electrical potential acquired from a muscular unit can be plotted and considered as a vital signal, called Electromyograph. Figure 1.2 depicts a typical electromyograph.

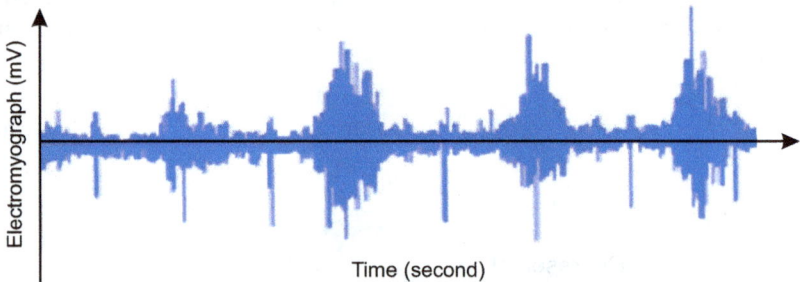

FIGURE 1.2: A rough representation of electromyograph.

Temporal profile of fluid pressure is also a signal, that is regarded as a vital signal if the fluid is blood. Regardless of the type of signal, there is always a variable of time associated with a variable of value representing the intensity of the phenomenon represented by the signal, or one can conclude that a signal is the variation profile of a phenomenon over time. Several related publications exist in the literature that categorise images either as signal or as data [105]. They introduce the processing commonly for images and signals, where an image is treated as a two-dimensional signal. In contradiction, this book considers image a separate class, which is not dealt with in terms of the methodologies. There are several motivations for this categorisation. Although some of the processing methods are essentially common for images and signals, a large number of the methods have been sophisticatedly introduced for signals only, in terms of both theoretical foundation and the implementation details. Furthermore, each pixel of an image may, or may not, have a physical dimension, depending on the image contents. Nevertheless, it is not decisive that a physical dimension is always associated with an image. On the other hand, an image is always captured or registered, like as a signal, and consequently cannot be classified as data. Deep learning methods for time series of the image is beyond the scope of this book and will not be addressed in the sequels of the book.

1.3 Data Versus Signal

The first difference that one can realise after the above descriptions, is an association of time with signal, in contrast with data where this association is not necessitated. The signal is often recorded in a previously known time interval, but data is not necessarily collected in a certain time sequence. For example, the facial colour of a baby might be collected a few hours after the birth, however, not all newborn babies must be attributed by their facial colour in some of the clinical routines. In contrary, signal is a registration in time with a certain order of time sequence. The data is, therefore, recorded or collected, whereas the signal is merely recorded. Another discriminating aspect of signal against data, is the dimension of signal. To clarify this point, a signal always corresponds to a phenomenon, represented by using a measuring technique, that provides a link between the phenomenon and the measuring technique with a known physical dimension. Data is not necessarily associated with a physical dimension, as not for the facial colour of a newborn baby. Moreover, the link to the physical phenomenon is not generally seen for data, but is so for the signal clearly. Sometimes data is a result of applying a processing algorithm on a signal and the resulting features extracted from the signal, constitute multi-dimensional data, from which there might be features that are not physically interpretative. Such data is often obtained by introducing a mathematical, or even a statistical mapping, applied to a set of certain types of signals. You will see the case studies in the following sections, in particular when the time series analysis is described.

1.4 Signal Models

Scientific studies toward understanding natural phenomena are mostly based on the pertinent models, capable of justifying different behaviors of the phenomena. A good model is the one that can explore different actions, interactions and also behaviors of the phenomenon. It is important to emphasize the fact that a model is indeed a way that we see a phenomenon, by assuming a set of the presumptions along with the range of the model applicability. The phenomenon by itself might show other sides in different conditions, which were not foreseen by the model. In general, there are three ways to model signals: deterministic, chaotic, and stochastic. Although similarities are seen in the processing methodologies, fundamental differences in the theories make such the classification necessary. In deterministic models, a signal is modelled by a closed mathematical formula, implying that behavior of the signal is well-recognized, and described by the formula. It is obvious that if a signal is completely modelled by a closed mathematical formula, then the value of the signal for the future as well as the past time, can be accurately predicted using the formula, which is clearly unpractical in real-world scenarios. In fact, deterministic models mostly correspond to the absolute definitions, and their applicability is mostly limited only to opening up a theoretical foundation. In deterministic methodologies, parameters like amplitude, phase, energy, and frequency are of importance for the processing. There is always a gap between deterministic models and applied solutions to respond practical questions, caused by variation of the signal behavior in time, called non-stationary behaviour of the signal, and also signal variation over subjects. This gap is mainly covered by further expansion of the deterministic models for specific practical questions, in a mathematical and sometimes statistical manner. Stochastic models come to practice when statistical methods are employed for modeling the signals. Stochastic models are based on the fact that amplitude of the signal is not strictly known in time, and is considered a random variable. Stochastic signals are attributed by the statistical parameters, like average and variance of the signals, in addition to those for the deterministic models. Many practical questions are better resolved thanks to the stochastic models. Chaotic models are basically deterministic with an initial value which is a random variable. Such models are sometimes called, disordered deterministic.

1.5   Noise and Interference

Noise is an unwanted, random-valued content, affecting a signal, with different unknown sources. So, there are always uncertainties associated with noise. Nevertheless, a kind of categorization is attributed to noise in sense of the possible source, to allow the engineers and researchers to justify behaviour of random processes appearing in the studies, and to select an appropriate statistical method to model the process. Thermal noise is the most common type of noise, seen in almost all electronic circuits. It results from any motion of the electrical charge carrier inside a semiconductor part of an electronic component. This is a random process the unavoidable occurs when the temperature of the component is above absolute zero. Thermal noise exits in a semiconductor even if no external electrical potential is applied. Thermal noise is statistically modelled by a Gaussian probability density function. Semiconductors are often deemed as a source of other kinds of the noise including flicker, shot, burst, and transient-time noise, which are mostly characterised by a white noise process. The common feature of all the sources of noise, is that noise introduces undesired contents to the signal, and does not correspond to any known physical phenomenon. Interference on contrary, is defined as the effect of an external signal (carrying information about a phenomenon), on another signal, by induction or any other coupling media. In the digital signal domain, other noise sources are accompanied, in addition to the above-mentioned sources. Quantization error is a typical characteristic of the digital signals, that inherently exist in the digital systems, which appears as a source of the noise [104]. Other sources of disturbance exist merely in the digital electronic circuits, like clock inference, acting as noise. Discussion about various behaviours of noise and interference is beyond the scope of this book.

1.6 Time Series Definition

A time series is a sequence of data points, collected in a priory known time order. The time order is not always, but often, expected to be equally spaced in time, meaning that the data point is sampled uniformly. A signal by itself is considered as time series, but a time series might be data points of an unknown phenomenon. The scope of time series, is therefore, broader than the digital signals, and covers the data partly. Time series can be multidimensional, resulting from applying a time-variant mathematical mapping to a signal. Time series of images is another important domain of study, which plays a role in different contexts, i.e., healthcare. Many biological signals, natural phenomena, and industrial activities are represented by the time series of recordings. As was mentioned, a digital signal is regarded as a one-dimensional time series, consequently, there is always a level of the random behaviour associated with the time series because of the noise existing in digital signals. In many applications, a time series of digital signals is mapped to another time series of multi-dimensional with different length, after feature extraction. Variation of a time series conveys information about dynamic contents of the time series, which can be exploited for different purposes of the learning process. This will be discussed in more detail in the related sections.

1.7   Time Series Analysis

Terrestrial phenomena can be generally registered and represented as the time series, based on the fact that earth is in an everlasting movement of the rotation as well as the turning, causing an oncoming movement of time that will never stop. Analysis of time series is performed for different purposes, mainly for five objectives: Classification, prediction, sequence detection, filtering, and simulation (or sometimes termed by generation). Each single point of a time series consists of information along with the noise, and extracting the information from various sources of contamination, is an essential task in any kind of time series analysis. Regardless of the objective of analysis, the main attention is paid to the dynamic contents of the analysing time series, in other words, exploiting information using the temporal variation of the data points. A time series is always contaminated by different sources of noise; therefore, statistical methods are often employed as the key-tool for the analysis, even though it might not be seen in the ultimate derivations. To put this point into a better perspective, sometimes time series analysis leads to the mathematical closed formulas for the analysis, which are not seemed to be statistical. Even though the statistical process might be unclear by the closed formula, the underlying model that provides the corresponding theoretical foundation often involves different statistical assumptions contributing to the method derivations. In most of the time series analysis towards classification and sequence detection, statistical methods are considered in two different directions; in learning dynamic contents of a single time series, and in learning data dynamic over different time series. In this perspective, an ergodic process is defined, as the process in which the statistical momentums are consistent over the ensembles and for a time slot of single time series. These points will be discussed in more detail in the related sections. Another point worth noticing is that adaptive methods are mainly applied to a single time series for the purpose of filtering or sometimes prediction, whereas the classification which is mostly performed using different groups of time series. These important notions will be further expanded in the corresponding sequels.

1.8   Deep Learning and Time Series Analysis

One important question that a reader can bring to mind, is the definition of learning, deep learning in an objective way. Although these terms will be discussed in detail, in the related chapters, a short introduction about the topic can be of help for better orientation in following to the corresponding contents. Learning by itself is the process of classification based on the similarities and dissimilarities. What brings difficulties to the learning, is indeed, the process through which the similarities and dissimilarities are found. In the conventional forms of learning, sometimes called shallow learning, one stage of processing is applied to the input data. This stage serves as the body process of learning, nevertheless, two other levels of processing are often invoked, named pre-processing and post-processing, which do not take part in the learning process, but instead provide a better condition for the learning. However, a deep learning process performs different stages of learning on the same input data, to find the similarities and dissimilarities of multi scale architecture. Deep learning methods are often employed either for classification of a time series using parts or all of the data samples, or for detecting certain sequences over a time series using short time samples. Much less attention has been paid to the other possible applications.

1.9 Organisation of the Book

This book provides principles for the students, researchers, and scientists in computer sciences, to find new ideas in learning and also to enable the researchers to express their ideas in an objective mathematical way. The book is organized in three parts, altogether comprising 12 sections. The first part provides introductory content about learning theory and validation in two chapters. A recently proposed method for validation, named A-Test, is introduced in Chapter 1 and recommended to be studied. This method elaborates the conventional validation methods, especially when it comes to the small data. Chapter 2, contains well-known fundamentals of learning. Part 2, brings the essentials of time series analysis with details of adaptation and classification. This part consists of three chapters, through which stochastic and deterministic theories will be further expanded. Recent theories, proposed for a group of time series, named cyclic time series will be deeply described and applicative contents of such an important group of time series will be explained. The readers are highly encouraged to pay attention to this chapter, as new ideas are specifically included in this book only. It founds a base for Chapter 7 and Chapter 8, with practical importance that can be found merely in this book. This part will be ended with an especial chapter on the dynamic learning method. For the experienced readers, this chapter can be skipped, however, mathematical derivations can provide means for the beginners to learn the principles. The main focus of the book is included in the third part. Chapter 6, explains the clustering methods which have been recently used for learning at the deep level. This is an elaborated notion of deep learning that can be found in this book, and differs from the existing ideas of deep learning. Chapter 7, gives materials on deep learning based on neural networks. Time growing neural network will be much expanded for the deep learning. Chapter 8, is sophisticated for the cyclic time series. This new idea facilitates learning in different practical application, especially in biology where cyclic time series are profoundly seen. Recurrent neural networks and convolutional neural networks are the material of Chapters 11 and 12. Experienced readers might be familiar with these chapters and can skip them. The book discusses hybrid methods for learning and their advantages and disadvantages against neural networks.

2

Learning Theory

Learning in its broad sense implies a process, through which similarities and dissimilarities of various classes of previously prepared data, sometimes called training data, are found, and an appropriate decision is made regarding it's implementation. In the context of computer sciences, learning is performed through a systematic procedure, known as the learning process. Here one should pay attention to the methodological differences in learning process between the conventional methods and the more recent forms of deep learning methods, especially when it comes to the time series analysis. In most of the conventional methods, a level of processing is involved, in which a mathematical or sometimes a statistical mapping is applied to an input time series. The result of the mapping is another time series, often a multidimensional time series, but with the shorter length in another domain of the numbers, which can be learnt in an easier way. This level of the process is called "feature extraction", which is expected to provide a mapping to another domain, in which the members with a similar class are closer in the same space. Feature extraction is an important part of a learning method, employed for quantifying similarities or dissimilarities over the samples of a dataset. In many of the examples of deep learning methods, this level of feature extraction is merged into the learning process, so more the process is executed the less hand-made manipulation is allowed. A learning process in the conventional form, is typically based on a set of the features extracted from the training data. Capability of one or a set of the features in segregating different classes of data, is termed as "discrimination power". Discrimination power is not an absolute definition. It is in contrast, employed to facilitate comparison of different feature sets. Nevertheless, the dependability of the discrimination power is not limited to the feature extraction only, while an underlying learning method holistically takes part in the value of the discrimination power. However, for comparing different feature sets, discrimination power can be potentially invoked when the rest of the learning process remains identical!

Another definition, worth mentioning at this point, is the terms, "supervised" and "unsupervised" learning or classification. Supervised learning, is a kind of learning in which classes of the members of the training data is priori known and employed for the learning process. An example of the supervised learning is classification of the Electrocardiogram (ECG) signals, where the label of the signals are clinically assigned by the physicians, and the learning method is deemed to find the similarity of the signals with similar classes, along with the dissimilarities of different classes. Unsupervised classification is however, the case in which these classes are not either known, or employed for the learning process, and the learning is performed based on the similarities only. By convention in this book, the labels of a supervised learning are denoted by capital letters. In the following sections, further definitions of learning are explained, sometimes by mathematical expressions, in conjunction with the principals needed for the following chapters.

2.1   Learning and Adaptation

In the first chapter, adaptation was addressed as an attribution of practical systems in particular, those inherently seen in all the elements of the nature as an everlasting change. Learning was discussed in contrast with adaptation, in which decision making takes a role in the process. However, indicative differences and similarities must be discussed in more detail, because of their important roles in selecting an appropriate strategy for solving a practical question. The border between learning and adaptation is sometimes unclear for many researchers, leading to a waste of time in developing improper solutions that cannot guaranty their research objectivity. It is important for an engineer or researcher to obtain an understanding about the intrinsic nature of the problem, before choosing a strategy for solving the research questions. In both the cases, a set of the parameters are invoked to provide a reliable result either for learning, or for the adaptation question. The number of the parameters associated with a practical system often resembles the order of the system. These parameters are dynamically estimated sometimes using a non-linear model to estimate the performance, however, as you will see in the following sections, there are often other parameters associated with this method, referred to as design parameters, which are found throughout the optimization process. Adaptive solutions are mostly employed when the dynamic of a phenomenon is taken into account. Adaptive filtering, Kaman filtering, and many other methods that have already been proposed in the adaptive control domain are typical examples of such solutions [30][2]. In contrary, learning can also be used with static data, when the data variation comes from the variation over the instances. The parameters of an adaptive method are updated by each input, which is not often the case for a learning method where the updates happen in the training phase. Decision making is also considered as another point of differentiation, often seen in learning methods, and not in the adaptive methods. Another point of divergence lays in the structure of the adaptive methods. A necessity of majority of the adaptive methods, is a reference time series with the informative contents, either coming from another channel of time series recording, or from delay of the same channel. Attention should be paid to the point that the reference time series is not necessarily a signal. It can be an interference, noise or even data. A line canceller which receives a power line from one channel and removes it from a signal contaminated by the power line interference, is a well-known example of such an application. Another example, is an adaptive controller for predicting body movement of an animal. There are also important differences in the nature of the two processes; adaptation and learning, making the scientists categorizing them into two different contexts, signal processing (or sometimes control) and machine learning, respectively. Adaptive methods have been widely presented in various text books of signal processing and control [22]. Although one strong suggestion for time series analysis can be adaptive methods such as the Linear Prediction Error Method, this book is planned to keep its focus on learning methods for time series analysis, and even more specifically, deep learning methods.

2.2 Learning in a Practical Example

Let's assume a bicycle trainee who is starting to learn bicycle riding. Visual, auditory, and motor skills of the person are all involved in sending signals to the bicycle rider's brain, which is responsible for receiving the signals, and to send suitable control commands to establish balance and to maintain appropriate movements. The learning process in this case includes, sending appropriate signals to the muscular units of rider, based on the visual, auditory, and muscular feedback. The rider should experience different situations of riding, in order to be able to maintain a stable riding experience. The rider's brain receives different sets of input data, which are in this case, visual, auditory and muscular feedback, and also performs a processing signal, and sends the commands to the muscular units, and finally validates performance of the bicycle rider based on the outcomes. In this process, a set of the input data with known consequences, resulting from previous experiences, plays an important role in the learning process. This set of data, is calling training data Maintaining the balance of the rider and providing an appropriate movement are accounted as the measures for the learning quality, which are termed as the "learning function". The input data are transferred to the brain, and those units of the brain responsible of receiving and interpreting the related data, take appropriate action. A massive interconnection of the synaptic links are made between the neurons in time when the rider experiences different conditions, and the rest of the brain cells remain almost neutral to the inputs. The number and distribution of the neural connection are parameters of the learning process that depend on the learning task, and also the input data. They are considered as the learning parameter. In this simple example, another set of the parameters contributing to the learning process, which seems to be hidden, but was set before the onset of the process. The number of the brain and muscular units, involved in the learning, and perhaps many other phenomenon, which are yet undiscovered, are typical examples of these parameters. These sets of the parameters, are named hyperparameters.

2.3   Mathematical View to Learning

Learning process is indeed the process of finding parameters of an appropriate mathematical mapping by which an input vector is mapped to another space where similarities and dissimilarities of the "within-class" and "between-class" data, are better projected, respectively. Figure 2.1 illustrates how this mathematical mapping functions.

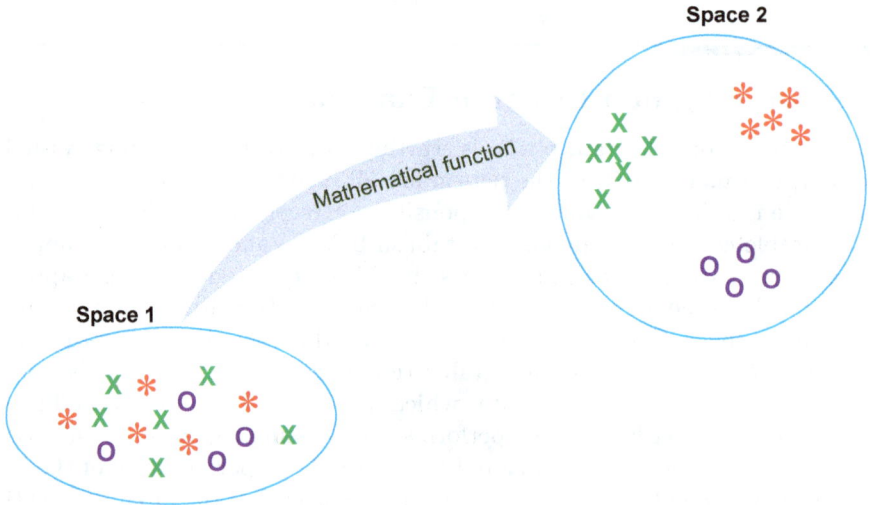

FIGURE 2.1: A suitable mathematical function can map the data from a 2 dimensional space to another space of 2 dimension, where the three possible classes are well segregated with high between-class, and low within-class variances.

In analogy, the term "Green Function", used in mathematics implying a process for finding parameters of a linear differential equation, with this difference that learning can include non-linear mappings as well. Depending on the learning fashion, a number of the attributions are associated with learning process, however, there are essentially some of the fundamental components, seen in all the learning processes. For the developers and students of artificial intelligence context, it is important to gain a deep understanding of the learning process, along with the learning objectives, to be able to achieve a reliable learning. Nowadays, learning methods have been massively developed and a bunch of the pertinent implemented codes can be publicly found. Nevertheless, finding the correct strategy to solve practical questions, and to develop individual solutions requires deep insight into the context, that begins by "understanding heart of learning process". Fundamental components of a learning process are described in the following sections.

2.3.1 Training and Validation Data

In the context of artificial intelligence, a predefined set of data is used to find learning parameters of a method which is called "training data". Looking back at the bicycle rider example described in the previous sections, the training data acts as the rider's experiences. Training data can be a set of the time series (in this case dynamic contents of the data are of importance), or single point registration of a phenomenon. In supervised learning, the class of each training data is known in advance. If the classes are assigned to all the training data instances using a set of the categorical data, the dataset is regarded as a fully labeled training data [81]. The labels are sometimes assigned independently, without looking at the signals by themselves, according to a priori understanding obtained from a different source of knowledge. An example of such supervised learning, is the case in which a set of the recordings of heart activities is deemed to be used for detecting heart conditions [51][46]. The recording labels are obtained by an expert physician based on clinical findings and other medical examinations. This example will be further explained throughout the book. In some cases, the labels are assigned at certain temporal intervals of the training time series, where trained eyes can detect several temporal sequences. Validation data is a set of the labelled data used for evaluation of the learning performance. Although validation data can be fully-labelled, the learning parameters are not updated by the validation data. In practical situations, we may encounter a problem in which several methods are to be trained and validated and we choose the one with optimal performance. The learning parameters of the methods are firstly obtained during the training which is followed by the optimization. Validation data is then used for determining the classification method offering the optimal performance. Sometimes validation data is invoked to select a suitable method such as the middle-classification method among several options. This is especially seen in the hybrid methods where a few classification methods are cascaded to deliver better classification performance [40]. This is the main difference between validation and training data that will be explained in more detail in the coming sections.

2.3.2 Training Method

The mathematical process, which provides the above-mentioned mapping, described as the training process, is known as the learning method. This can rely on a series of mathematical functions, even though a single function which is capable to serve as such the mapper can be considered as a learning method [24]. In many cases, the learning method is corresponded to an estimation, in the sense of the mathematical action. To bring this point into a better perspective, let's consider the case in which a linear system of equation is supposed to be solved. In order to achieve a unique solution for a system of equations, the following criteria must be fulfilled:

- The number of the equations, N, must be equal to the number of the variables M

- The equations must be linearly independent

A curious reader may come up with this question: what happens if the first criterion is not satisfied? In this case, we might be faced with two situations;

- The number of the equations is lower than the number of the variables

- The number of the equations is higher than the number of the variables

If the number of the equations is lower than the number of the variables, and the equations are linearly independent, a unique solution does not exist, and there will be unlimited number of the solutions. The most important part is, however, the situation in which the number of the equations is higher than the number of the variables. In this case, the system of equations cannot derive any unique solution, and this leads to "estimation" question. This point is summarized in Figure 2.2:

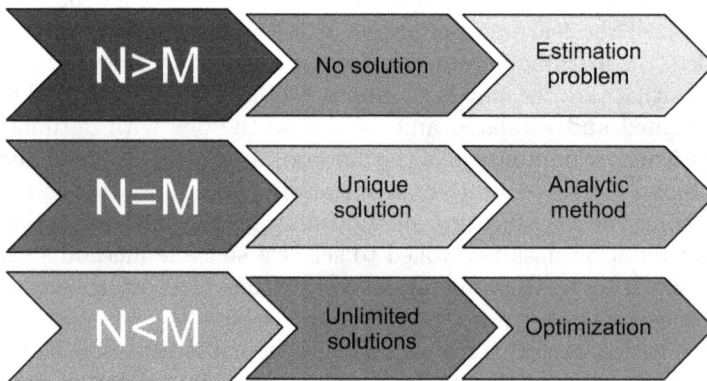

N>M	No solution	Estimation problem
N=M	Unique solution	Analytic method
N<M	Unlimited solutions	Optimization

FIGURE 2.2: All the possible situations that may happen for a equation system, in terms of N and M, defined as the number of the equation and the number of the variables, respectively.

An estimation process often consists of a procedural method in conjunction with a set of the criteria that should be optimized through the estimation procedure. This set of the criteria is sometimes called "cost function". The learning method, in its intrinsic form, performs an estimation, since we have a set of data to be mapped, to several classes and the set of data is always higher than the number of the classes. Looking back at the example of the bicycle rider, the number of the instances sent to the brain of the rider is by far higher than the conditions. In most of the cases of learning, the estimation is performed through an iterative procedure and the cost function is calculated as the mean square error of the predicted and actual values. This will be further expanded with more detail in the upcoming sections. Artificial neural network, discriminant analysis, and fuzzy logic are considered as the three well-known learning methods, that will be partly explained in the related sections.

2.3.3 Training Parameters

In artificial intelligence, a learning process is often initiated by following a certain procedure, described by the learning method, in a way to optimize a criterion such as minimizing a cost function. The criterion, can be the mean square error resulted from the subtraction of the desired class and the predicted class of each data sample. Another criterion for the learning, is for example: the quantified similarities over the data. Sometimes, the optimization is performed through a constrained minimization, as done in a support vector machine [141]. In many cases, the learning process is performed through an iterative procedure where numerical analysis is widely employed. As a result, a number of the parameters are obtained through the learning procedure, most of the times iteratively (especially in the elaborated methods), but sometimes analytically by invoking techniques and lemmas from linear algebra [141]. Even the analytical solutions need iterative methods from numerical analysis for the implementation. In any case, the obtained learning parameters can be used to perform the mathematical mapping, which makes extraction of the similarities and dissimilarities feasible. It is important to note that the learning parameters are all obtained for a specific solution with a certain training data. The learning parameters are tightly linked to the training data, and updated at each iteration of the learning process. This type of the straight data-dependent feature of the learning parameters, with updating by each training data, is a certain feature of the parameter that makes them different from another set of the parameters, named hyperparameters.

2.3.4 Hyperparameters

Hyperparameters are those parameters which are set before starting the training phase. These parameters are sometimes called, design parameters, however, hyperparameter is the term that is currently more popular in the context of artificial intelligence. From the example of the bicycle trainer, one can clearly see that the neural links and synaptic connections, made during the training, are considered as the learning parameters. However, the signals received from the rider, were initially set prior to the training task. The signals, are visual, auditory and muscular feedback. The number of the signals, and the action of the signals in the learning process, are initially set (unintentionally in a biological system), before the training, resembling the design parameters.

2.4   Learning Phases

From the previous explanations, it can be easily seen that a learning process typically involves five different phases of activities: data preparation, problem formulation, training, validation and optimization. Figure 2.3 illustrates a typical learning process:

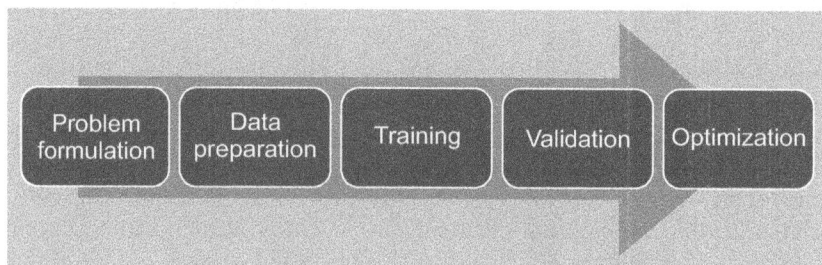

FIGURE 2.3: Typical steps of a learning process.

The first phase of learning is often assigned to finding appropriate data for the learning process. Validity of data is an important task that must be performed prior to any process. The data must be valid, informative and clean, otherwise efficient learning cannot be expected. Result of an inefficient learning may be seen as an incorrect classification for different input samples. Depending on the data type, an appropriate strategy must be selected to purify the training dataset. As an example, the learning process should probably take dynamic contents of data into account, when it comes with the time series classification, while for a development problem of face detection, dynamic contents of the data are not available. Selecting an appropriate learning strategy helps to provide a clear formulation of the problem. The problem formulation phase comprises of selecting the learning strategy, identification of the hyperparameters and finding the learning parameters. Training phase includes choosing a training method, and implementing the training procedure. The training phase leads to a set of the parameters, having the hyperparameters already initialised. It is worth noting that a set of the hyperparameters is selected before the onset of the training, either intuitively by having a prior knowledge of the case study, or by following a systematic procedure. After the training process, the trained system is validated using a set of the input data, normally out of the training data, but in some cases combined with the training data.

The result of the validation indicates suitability of the learning performance. Optimization is a process through which the hyperparameters are found, that results in an optimal performance. In many development problems, due to the insufficient theoretical bases for finding the hyperparameters in an analytic way, systematic procedures are followed to obtain the optimized parameters. Many developers offer the use of the classification rate as a metric for their optimization. Depending on the case study, there are other metrics that can be invoked for the optimization. The metrics which profile the performance measures will be described in the following sections.

2.5 Training, Validation, and Test

Training refers to a process for finding learning parameters of a classifier. In most of the cases the learning parameters are incrementally adjusted through a recursive procedure in a way to minimize a cost function at each recursion of training. Regardless of using recursive training procedure with the numerical analysis or using an analytic solution to find the parameters of the classifier, a training method is always associated with the training procedure, that governs the procedure on how to find the learning parameters. In most of the cases, the procedure is repeated several times with a certain dataset, named the training data, until reaching a certain low level of the cost function. In some of the classification methods, the recursion can be continued until reaching a very low value, ideally zero value, of the cost function with a certain training data [81]. An important issue, however, arises when the classifier is experimented on by another set of the data out of the training data. A classifier can be well-trained with a very low value of the cost function, but show a high error in the validation. This is considered as a poor performance for the real world practices. It is therefore necessary to examine the classifier using a different set of the data, otherwise the training would be unreliable. The data which is used to validate performance of a classifier is called the validation data. In practice, there are various situations where the data is not included in the training data. This necessitates to validate a classification method using a dataset, different from the training data. It is also practically seen that there exist several alternatives of classification method on the table to be selected, for the same classification purpose. The main purpose of using validation data is to chose the classifier with the optimal performance. The training data is conclusively employed to find the learning parameters, and/or hyper parameters of a classifier, while the validation data is used for selecting an optimal classification method. A classifier, after being trained and validated by the training and validation data, is evaluated by another set of data, named the test data. The notation of N_t, N_v, and N_e are the notations will be used to denote, the number of the samples for training, validation and test, respectively, throughout all the chapters of this book. It is obvious that the total number of the samples is a summation of all the samples:

$$N = N_t + N_v + N_e \qquad (2.1)$$

In some of the application, the learning dataset, N_l, is regarded as sum of the training and validation data while the test data remains unknown in the learning phase:

$$N_l = N_t + N_v \qquad (2.2)$$

It is sometimes seen that the learning dataset is not defined and the processing suffices to the training, validation and test only.

2.6   Learning Schemes

In all the learning questions, depending on the input dataset and the characteristics of the data, an appropriate model is chosen for the learning. The model chosen for the learning, is completely tied to the type of the data that should be learnt, in terms of its nature and also availability of the data label. Basically, one can face two different cases of learning: supervised and unsupervised learning. Each of the cases, by itself includes two sub-cases: dynamic and static learning. The learning schemes corresponding to the cases are described in more detail in the following sequels:

2.6.1   Supervised-Static Learning

Supervised-Static learning is a learning case in which the training data is fully labelled and the training method does not consider temporal variation of the data in the architecture of the method. A preliminary requirement for any supervised-static learning is a set of the previously-known labels, $\{Q_i : i = 1, ..., N_l\}$, associated with the learning data, and the data samples are not time series of $\{x_i \in \mathbb{R}^n\}$. In most of the cases, the labels can be assigned as the numerical or categorical data or symbols, $(\forall i : Q_i \in \mathbb{Z})$, or a character $(\forall i : Q_i \in \{a, b, .., z\})$. A supervised-static learning, is a process which employs the labels for the training and for the validation.

2.6.2   Supervised-Dynamic Learning

In supervised-dynamic learning, each sample of the learning data consists of a time series: $\{\forall i : x_i \in \mathbb{R}^{n \times m}\}$, where n is the dimension, and m is the length of the time series. In many cases, the learning data contains time series of different length: $\{\forall i : x_i \in \mathbb{R}^{n \times m_i}\}$), and m_i is the length of the time series x_i. For the supervised-dynamic, we may encounter with the cases, where a single label, $\{Q_i : i = 1, \ldots, N_l\}$, ($N_l$ is the size of the learning data), is assigned to the entire of time series i. However, the labels are sometimes dynamic, with equal length equal to the time series samples: $\{Q_i(t) : i = 1, \ldots, N_l, t = 1, \ldots, m_i\}$. It is obvious that a supervised-dynamic learning method employs both the set of the labels and the time series for the learning process.

2.6.3   Unsupervised-Static Learning

Unsupervised-static learning is a process, in which the learning data is composed of a set of the vectors without any label. The process attempts to extract similarities disregarding any initial label, and terminates after assigning unique labels to all the vectors, $\{\forall i : q_i \in \mathbb{Z}$, to each sample of the training data $\{x_i : i = 1, \ldots, N_t\}$. The number of the classes is regarded as a hyperparameter. The learning process can be iterative and the labels may vary several times for each single sample throughout the learning process, before reaching an ultimate and stable set of the labels. Sometimes it doesn't converge to a stable set of the labels.

2.6.4   Unsupervised-Dynamic Learning

An unsupervised-dynamic learning is similar to the unsupervised-static one, with the difference of including dynamics of time series at the input. Quantifying the learning performance would be considerably different when it comes to the input data of time series.

2.7   Training Criteria

A dynamic process for extracting similarities and dissimilarities of the input data, termed by learning process, demands a systematic procedure towards achieving certain goals such as optimal learning quality. A question at this point on the learning process, is how to quantify the learning quality, using a reckoning learning data. This point directs us to the learning criteria, which addresses the way of quantifying the learning quality in a predefined situation. In fact, the learning scheme plays the key role in the quantification method. Classification error, defined as the ratio of the incorrectly classified samples to the total samples, is one of the most common criterion employed in majority of the supervised learning methods. Considering the following function:

$$\delta(x, y) = \begin{cases} 1, & \text{if } x = y \\ 0, & \text{if } x \neq y \end{cases} \tag{2.3}$$

Classification error, I_e, of a classifier, is defined as:

$$I_e = \frac{\sum_{i=1}^{N_v}(1 - \delta(Q_i, q_i))}{N_v} \tag{2.4}$$

where N_v is the number of the validation samples. Q_i is the actual label of the validation sample i, and q_i is the label resulted from the classifier. In most of the supervised learning, this criterion is attempted to be minimised during the training phase. The lower classification error results from the validation phase, the better training is achieved. In contrary, classification rate is defined as the ratio of the correctly classified samples to the total number of the validation samples, as follows:

$$I_R = \frac{\sum_{i=1}^{N_v} \delta(Q_i, q_i)}{N_v} \tag{2.5}$$

It is obviously seen:

$$I_R = 1 - I_e \tag{2.6}$$

These performance measures are sometimes used in percentage (%) form. In this case we obviously have:

$$I_R|_{\%} = 100 - I_e|_{\%} \tag{2.7}$$

Classification rate is commonly used as a performance measure for comparing different classification methods. In unsupervised learning, similarity between samples is regarded as a criterion for the learning, during the training processes. There are different ways to formulate the similarities. Normalized cross-correlation of the two vectors, $X_1 = [x_{1,1}, ..., x_{1,m}]^T$ and $X_2 = [x_{2,1}, ..., x_{2,m}]^T$, is one of the most common performance measures, which is calculated as follows:

$$\Gamma_{X_1, X_2} = \frac{X_1^T \cdot X_2}{\|X_1\| . \|X_2\|} \tag{2.8}$$

$$\|X_1\| = \sqrt{x_1^2 + ... + x_m^2}$$

If the two vectors are completely dissimilar, the corresponding cross-correlation approaches to -1, and in contrary, for two similar vectors this metrics becomes well close to 1. Distance measurement is another way of quantifying dissimilarity of two vectors. Euclidean distance is one of the most common distance measurement techniques:

$$d(X_1, X_2) = \sqrt{(x_{1,1} - x_{2,1})^2 + ... + (x_{1,m} - x_{2,m})^2} \tag{2.9}$$

Two vectors are dissimilar if their distance is large. Distance measurement is usually used for comparison since it lacks from a standard to scale similarity such as large and small. Unlike the normalized cross-correlation, where the similarities are normalized between the two limits, -1 and $+1$, standard values to quantify the comparative description, e.g., large, cannot be derived by distance measurement. Therefore, such a normalization depends on the case studies rather than objectively known values. This limits application of the distance measurement techniques only for the certain cases, in which the distance values are interpretive. By the way, the cross-correlation can be easily reformulated for measuring dissimilarities between the samples.

2.8   Optimization, Training, and Learning

Optimisation is a process through which **design parameters (or hperparameters)** of a learning method are found. Training process is, however, a process for finding the **training parameters (training weights)** instead of design parameters, by assuming that the design parameter were already assigned. The learning process in particular, incorporates optimization, training and validation, implying on the process that encompasses finding both the design parameters and the learning parameters. In an optimization process, a cost function is attempted to be minimized, either analytically using a mathematical derivation, or numerically through a pre-designed procedure mainly by a set of the iterative operations. In computer systems, it is always preferred to perform iterative procedures with low computation power, composed of simple arithmetic operations, instead of performing heavy mathematical calculations like obtaining inverse matrix, which might lead to a singularity. This will secure the operation of trapping in a singular point, and considerably improve the performance. As a result, in many cases, an analytical optimization method is tried to be converted to an iterative procedure, as mush as possible, even at the expense of losing accuracy in a modest way. The extent to which the accuracy is lost is obviously dependent to the computational power and the number of the iterations performed through the procedure. Sometimes an iterative procedure invokes experimental calculation for the optimization, as happens mostly in the intelligent methods. It is important to note that optimization is not limited to the intelligent methods only, and is generally employed in different fields, such as communication systems, adaptive filtering, and economic systems. In training, in contrast, a number of the criteria or constrains are optimized, mostly through a recursive procedure training is found merely in the intelligent methods. Constrained training is a common form of the training seen in several training methods such as support vector machine.

2.9   Evaluation of Learning Performance

An intelligent method, is always needed to be evaluated after a learning process, as like as a child who makes efforts to learn a skill and is evaluated by the trainer after the learning process how the learning goals are met. The first step in any assessment or evaluation process is quantification of the learning goals. The following sections address a number of the most common methods for evaluation and comparison of classifiers.

2.9.1   Structural Risk

In artificial intelligence, a clear understanding about performance of classification methods is sometimes a big challenge.

Structural risk of a classification method is termed as instability of the method in its performance measures, when the method is evaluated by a dataset out of the training data [44][43].

High structural risk is undesirable in any classification method. A well-known performance measure of a classification method, is classification error (see Eq. 2.4) of the classification method. Depending on the classification purpose and the case study, other performance measures like *sensitivity* might be of interest to be considered for the evaluation. This is seen in medical applications where an abnormal condition is learned against a normal condition and defined as the percentage of the abnormal cases which are correctly classified by the method. In many practical situations, especially in the problems involving supervised learning, it is commonly seen that a certain classifier learns similarities of the training data and the classification performance is tightly dependant on the training data. Such the similarities might not be seen in a test data, and therefore, accuracy of a method is not stable when the method is going to be tested by different sets of the data out of the training data. A common condition which can lead to a high structural risk, is a problem called over-fitting. This happens when the classifier becomes over sensitive to the similarities such that even small unimportant dissimilarities are learned by the classifier. This point will be further discussed in the coming sections. One must bear in mind that although training data plays an important role in the learning process, structural risk addresses a deficit in the architecture of the classifier which makes most of its effort to extract unimportant similarities and dissimilarities. A classifier with high structural risk is not reliable and can give an unacceptable classification rate, varying with different testing datasets, in practical situations.

2.9.2   Empirical Risk

Classification error of a classifier, as defined in Eq. 2.4, is sometimes calculated by using a finite set of the training data, and used as a crude measure of the possible error in reality. It is obvious that the performance is mostly over-estimated during the validation. Choosing an appropriate validation method can push the estimated performance to the actual ones in terms of the classification error. The empirical risk, although is crude, but frequently used in many statistical classification methods, especially with large training dataset. A common condition which results in an unrealistic estimation of classification rate is a condition in which the training process is stuck in a situation, named local minimum. This condition can happen in training of a neural network using conventional back propagation error. Convex classifiers like support vector machine have this interesting aspect of preventing occurrence of this condition.

2.9.3 Overfitting and Underfitting Risk

Machine learning methods, in a global sense, are divided into two different categories in terms of the learning fashion, even though objective of the two categories seems to be identical: convex and locally optimized methods. The former refer to the methods, where the learning process ends up to global optimal point of the cost function, whereas the later one in which the optimal point cannot be globally guaranteed. Support vector machine and neural networks are typical examples of the convex and locally optimized methods, respectively. Learning parameters of a convex classifier are mostly calculated by using closed analytic formulas. Even though, iterative procedures are sometimes preferred in order to decrease the computational complexities or to run away from the singularities, the global optimum is theoretically known. In contrary, for classifiers of the second category, the global optimum is not known, but tried to be achieved through a recursive procedure. One might think that the more recursion to be employed, a closer point to the global optimum is achieved. This is true only in theory. In practice, any classifier whose learning parameters are obtained through a recursive procedure, can be faced with two different risks affecting performance of the classifier: overfitting and underfitting. Overfitting is a condition in which the number of the recursions has been unreasonably increased such that the decision border between different classes is very tightly set. Figure 2.4 shows a case of the overfitting. On the other side, there might be learning cases where the number of the recursions is insufficient in a way that the border is coarsely set, as shown in Figure 2.4.

Both the overfitting and underfitting bring risks to the classifier. The former might show low empirical risk at the expense of high structural risk, while the vice versa is for the later, with higher empirical risk, but probably a better structural risk. There is no optimal point in between, to be found analytically, and the trade-off between the risks must be performed through a process, so called "cross-validation". By definition, cross validation implies on a systematic interrupting of the learning process at the training level (which is assumed to be recursive in this case), and evaluating the performance using the validation data, and repeating this procedure with a sufficient number of the iteration until reaching a stable result.

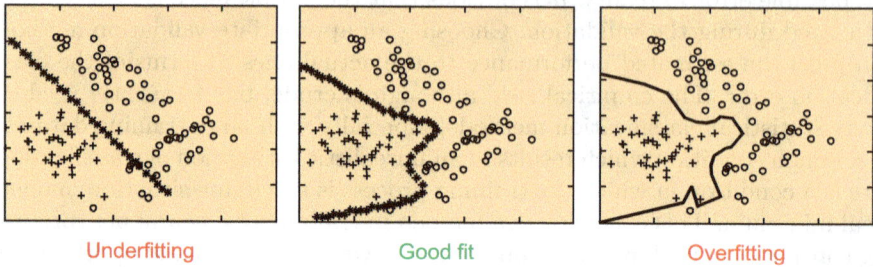

Underfitting   Good fit   Overfitting

FIGURE 2.4: The risks of overfitting, underfitting against a good fit, that can affect the learning process.

2.9.4 Learning Capacity

The ability of a classifier in improving its learning performance with more training data, is defined as the learning capacity. This lays well within the context of artificial intelligence which pays more attention to learning quality. One of the most common ways to quantify the learning performance, and hence to provide an understanding about the learning quality, is classification rate (see Eq. 2.5). This metric is calculated not only during the training, but also after the learning process, for estimating performance of classifiers and also for comparing performance of different classifiers. Although the objectivity of this criterion has been well-received so far, especially when a certain common learning data is employed for validation of different classifiers, there are other sides to this point which are worth discussing.

Learning performance can be explored in two different perspectives: during training and within validation. The trend of improvement in the classification rate during the training phase shows how well a classifier performs its task, which can be assumed as an indication of the learning capacity. A quick learner is always of interest, especially when it comes with a large training dataset. Another perspective of learning capacity, which is by far more important, is trend of improvement in the classification rate after validation with a different size of the training data. Almost in all the applications in the context of machine learning, the training data is cumulatively increased in time, and it is important to obtain an understanding of how the classifiers behave when they have been trained by a larger selection of training data. This aspect of the learning capacity is crucially important and regarded as an objective quality of classifiers.

2.10 Validation

In many practical situations, it is seen that a well-trained classifier gives a poor result in terms of the classification rate after been tested by test data, mostly due to the inappropriate training. In fact, it is always difficult to obtain a realistic estimation of the classification performance of a classifier, however, by following an appropriate validation, one can simulate the real situation. The extent to which the simulation imitates a real situation depends on pervasiveness of the training data, especially in a small or medium training data size, It is important for training data to include samples of every possible situation. Consequently, one cannot easily judge how a previously trained classifier behaves with a new input which was not trained for. That's why the validation method becomes important.

A number of the validation methods which are commonly used in artificial intelligence domain for validating classifiers are explained in this section. It is important to note that all the below described methods are often used to obtain an understanding about the structural risk of a classification method. This is indeed a simulation of the real situations rather than standardization of the risks. A difficulty in such analysis, commonly seen in all the three following methods, is their dependence to the learning data, since one may not know the true statistical distribution of the data. It is therefore, of critical importance for learning data to pervasively cover various conditions of the data. This can be difficult in many practical situations.

2.10.1   Repeated Random Sub Sampling (RRSS)

In this method, a certain portion of the learning data is selected for training a classifier and the rest of the data is used for calculating the classification performance [72][68]. The proportion pf the test data is fixed, but the data is randomly selected based on using a persumed statistical distribution (typically uniform distribution) and this procedure is repeated several times, with the random selection of the training data. However, size of the training data is fixed, and statistical descriptive of the classification rate is used to estimate the performance of the classifier and also for comparing different classification methods.

2.10.2   K-Fold Validation

This method is used for validating the performance of a classifier. In the K-Fold method, the whole learning data set is divided into K partitions of equal length. Then, one partition is used for validating the classifier and the rest of the data for training the classifier. This procedure is repeated K times with one partition used only once for the validation. This method is often preferred over repeated random sub sampling, when the size of the learning data is low. An exceptional case of K-Fold validation method, is the Leave-One-Out method, or alternatively, Jack-knife, method. In Leave-One-Out method, a single sample of the learning data is used for validation, and the rest for training the classifier. The procedure is repeated by the same number as the learning data size, with one sample in used only once for the validation.

2.10.3 A-Test Validation

A-Test method is indeed an elaboration of the two former ones, which can provide better information about the performance of classifiers [56]. This method is based on using K-Fold validation with different values of k. In this method, k is known as the validation index. Then, the learning data is randomly shuffled and the same procedure is repeated until reaching stable values for the classification rates obtained for each value of K. Figure 2.5 illustrates the method in its general form:

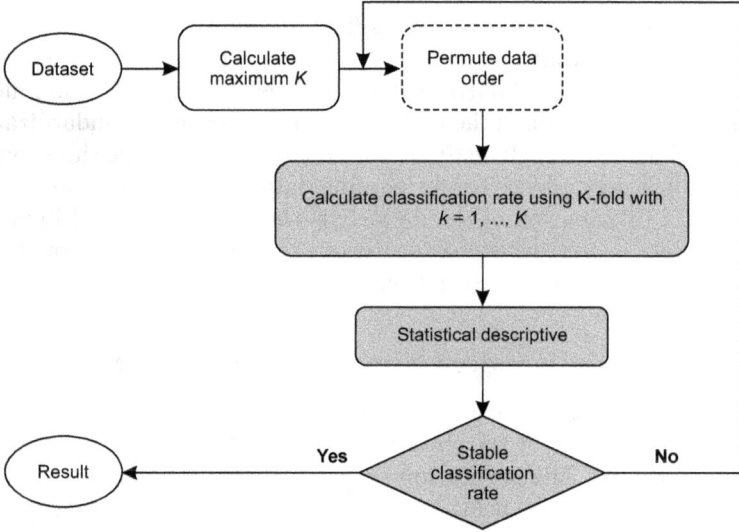

FIGURE 2.5: A complete block diagram of A-Test validation method. In order to expedite the process, permutation of the data might be ignored (the block denoted by the dashed line).

For each value of k, the classification rate is obtained and treated as a random variable. The probability expectation of this random variable provides an indication of the classification rate for a certain k. Let's I_R be classification rate of a classifier, the expected value of the classification rate for the validation index k is:

$$I_R(k) = E\left\{ \frac{\sum_{i=1}^{N_v} \delta(Q_i, q_i(k))}{N_v} \right\} \quad (2.10)$$

where is N_v number of the validation samples and $q_i(k)$ is the classification result for the sample i that is obtained using validation index of K. The upper limit of k is found by paying attention to the size of the validation data. It is obvious that the K-Fold validation method in its especial case, becomes Leave-One-Out, when the testing partition incorporates only 1 sample of the learning data. This condition happens if the k value exceeds $\lfloor \frac{N_v}{2} \rfloor + 1$, where the operator $\lfloor . \rfloor$ points to rounding a decimal number to its lower value. On the other hand, the validation index cannot take a value lower than 2, in which half of the learning data is used for the training and the rest of the other half for validation. The range of k is therefore expressed by:

$$k = 2, ..., \lfloor \frac{N_v}{2} \rfloor + 1 \tag{2.11}$$

Having the range of k, the classification rate resulting from A-Test method is:

$$\mathcal{I}_R = \frac{\sum_{k=2}^{K} \mathrm{I}_R(k)}{K-1}$$
$$K = \lfloor \frac{N_v}{2} \rfloor + 1 \tag{2.12}$$

The scuffling part of the A-Test guarantees that the learning data is not arranged in a specific order to affect a fair validation, and therefore risk of the bias on a certain dataset is considerably mitigated. Depending on the comprehensiveness of the learning data as well as distribution of the data classes, random arrangement of the validation data can be sometimes done only once before validation. This might lead to loosing the elaborateness of the method at the expense of the simplicity, even though, in many cases such estimation of the performance measure will be sufficient for validation and comparison. In this case, Eq. 2.10 is simplified as:

$$\mathrm{I}_R(k) = \frac{\sum_{i=1}^{N_v} \delta(Q_i, q_i(k))}{N_v}$$
$$\mathcal{I}_R = \frac{\sum_{k=2}^{K} \mathrm{I}_R(k)}{K-1} \tag{2.13}$$

As seen in the above equations, the A-Test method provides a set of the values projecting the performance measure for different validation index, k. Some indicative information can be extracted from this set of the performance measures, in different illustrative and quantitative ways. Advantages and disadvantages of this method against the other two methods, repeated random sub sampling and K-Fold validation, will be described in the following sections.

2.11 Privileges of A-Test Method

A-Test offers a more realistic, broader information and deeper understanding about performance of classifiers in reality as being compared to the other two validation alternatives. These privileges are obtained at the expense of more complexities in calculation. The following subsections brings up some of the interesting aspects of the A-Test method along with the informative contents which can be exploited by using A-Test validation.

2.11.1   A-Test and Structural Risk

In equation 2.10, the classification error was calculated for different values of K, not only for the validation, but also for comparison of different classifiers. The A-Test method can provide a measure about the structural risk of a classifier, if classification error, I_e, is invoked instead of the classification rate:

$$\mathrm{I}_{SR}(k) = \mathrm{E}\left\{ \frac{\sum_{i=1}^{N_v}(1 - \delta(Q_i, q_i(k)))}{N_v} \right\}$$
$$\mathcal{I}_{SR} = \frac{\sum_{k=2}^{K} \mathrm{I}_{SR}(k)}{K - 1} \tag{2.14}$$

where K is defined in (2.11).

A graphical representation of the classification error can provide an illustrative representation for different classification methods. Figure 2.6 demonstrates a typical graph for different classification methods.

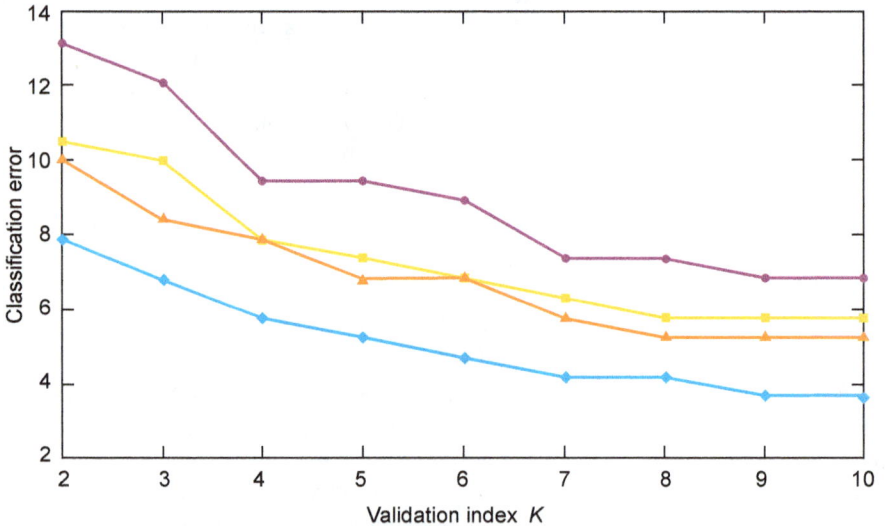

FIGURE 2.6: Illustrative comparison of 4 classification method suing A-Test method. A-Test greaph is depicted for four different neural networks: Time Growing neural network (blue), time-delayed neural network with two different windows length (yellow and orange) and multi layer neural network (violet).

As can be seen, the two time delayed neural networks show similar performance for some of the validation indices ($k = 4, 6$), nevertheless the orange one outperforms the yellow graph. As a result, a misleading comparison can be made if one suffices to 4-fold or 6-fold validation for comparison. The A-Test provides an informative illustration to compare different classification methods. Besides, the slope of the classification error with respect to the validation index, mean value and standard deviation of the classification error,

altogether exhibit how well the performance of a classifier is improved when
size of the training data is increased. Such the valuable information cannot be
obtained by the other methods.

2.11.2   A-Test and Leaning Capacity

Remembering from the previous sections that learning capacity of a classifier
is defined as the potential of improvement in the performance of the classifier
when the training data is enriched. In contrast with the other two methods,
A-Test undertakes the validation with different sizes of the training data, and
hence makes an estimation of the learning capacity feasible. One way to this
end, is to find a graph of variation of classification rate with respect to the
validation index. Slope of the graph along with the range of the variation can
provide valuable information about the learning capacity. One way to formulate the learning capacity is to calculate relative variation of the classification
rate, given in (2.12). The following derivation gives a simple way to estimate
learning capacity in percent (%) based on using A-Test method:

$$\mathcal{I}_{LC} = 100 \cdot \frac{\max \mathrm{I}_R(k) - \min \mathrm{I}_R(k)}{\min \mathrm{I}_R(k)} \tag{2.15}$$

$$(\min \mathrm{I}_R(k)) > 0$$

where $\mathrm{I}_R(k)$ is given by Eq. 2.10. It is obvious that the above derivation tends
to infinity when the minimum classification rate tends to zero. Such condition
creates singularity to the estimation of the learning capacity, showing that the
applicability of the method is limited to the engineering pursuits only.

A large $\mathcal{I}_{LC}$ for a classifier shows that the classifier can exhibit a much
better performance if being trained by a larger training data. In another word,
a classifier with high $\mathcal{I}_{LC}$, but not so high as $\mathcal{I}_R$, can be regarded as a potentially good classifier with a high capacity of improvement, even though it
is not practically the case for a small dataset. It can be predicted that the
classification error can be drastically decreased if being trained by a larger
training data. Such the interesting information cannot be provided by other
validation methods.

It is worth noting that in A-Test validation method, neither the absolute value, nor the average value of the classification rate can be employed as
the indicative qualifications for the learning capacity, instead variation of the
classification rate conveys the information about the learning capacity using a
larger training data. In many medical and clinical applications, preparing an
initial training data can be expensive, and sometimes impossible. Therefore,
choosing an optimal classification method among a great variety of the available ones, is critically important. Oppositely, in the testing phase, obtaining a
large group of the data from the referrals who would undergo the clinical test
can be by far more inexpensive. This is consequently, essential to calculate the

learning capacity of the classification method in the training phase to select an optimal method which will ensure superior long-run performance.

2.11.3   A-Test vs other Methods

Repeated random sub sampling is based on random selection of the training data, with a fixed size of the training data. Although uniform statistical distribution is often used for the random selection of training data, there is no guarantee to secure that each sample of the learning data receives equal share in the validation. In contrary, K-Fold validation assigns equal share to each data sample, to contribute in the validation. Nevertheless, repeated random sub sampling can provide a condition to calculate a confidence interval for classification rate, whereas K-Fold validation gives a unique value for the classification. A-Test on the other hand, can provide a confidence interval for classification rate, and meanwhile assigns a consistent contribution to validation for all the learning data samples, even if the shuffling part is excluded. Besides, the two important and unique possibilities offered by this method against the other two: estimation of the structural risk and learning capacity, make the A-Test method well preferred over the others, for most of the applications in artificial intelligence, especially when the learning data is not sufficiently large. Table 2.1 lists the indicative features of the three methods.

TABLE 2.1: Capabilities of the three validation methods commonly used in artificial intelligence; Repeated Random Sub Sampling (RRSS), K-Fold, and A-Test, in terms of the possibilities to render results of the performance measures for a learning method.

Validation Method	RRSS	K-Fold	A-Test
Calculation of descriptive statistics	✓	−	✓
Consistent share to all samples from validation data	−	✓	✓
Validation using various train/test ratios	−	−	✓
Intuition about the structural risks	✓	✓	✓
Quantitation of structural risk	−	−	✓
Calculation of learning capacity	−	−	✓

These possibilities cannot be provided by the other two methods where the size of the training data is fixed. A powerful strategy to permute the learning data in a way to guarantee consistent weight to each data sample is not a heavy task in many applications. Apart from the statistical methods several other methods can be developed to undertake this task. This is however beyond the scope of this book.

2.12   Large and Small Training Data

In practice, machine learning experts can find themselves in different challenges, when they develop machine learning methods with large or small training data. For a classifier, large training data is a condition in which the training data incorporates adequate samples of different possible classes. One can easily see that large is not an absolute term, and depends on the number of the design parameters. Here, it is important to note that there is a big difference between the topics "big data" and large training data. Big data, mostly corresponds to the problems in which we may have (not necessarily have) large training data, where each data sample by itself consists of a large data. An example, is recordings of heart signals from a patient group, where each patient recording contains several millions of data samples. During training with a classifier with large training data, it is important to maintain the training under a condition, where we uniformly include different samples such that certain data is not dominantly contributing in the training. Otherwise, the classifier will be clearly biased to learn that dataset is better than the rest. In contrary, for small data cases, we might not have adequate samples from each data form. The main problem in this case is overfitting and structural risk. This is the point where A-Test validation method is highly recommended to be used for the validation. Nevertheless, for large training data, an appropriate K-Fold validation can sometimes serve as a powerful validation method with an approximately realistic estimation of the classification rate.

3

Pre-processing and Visualisation

The time series of many physical phenomena in most cases carry uncountable information by a sequence of the data samples recorded in a certain temporal order. It is often important to find efficient methods in order to extract informative contents of a time series while preserving dynamics of the time series in a concise form to be used either for classification or for representation purposes [159]. Part of this process is known as feature extraction. Feature extraction is indeed a mapping of a time series of length $n \times m$ to another domain of multidimensional time series of length $k \times l$ in a way to provide a better segregation between different classes exist in the learning data. Nevertheless, the number of the features may be so large that another process needs to be invoked to reduce dimension of the feature vectors, and meanwhile to make the resulting time series further informative. This process is named pre-processing, which is sometimes employed for representation only. The pre-processing is sometimes employed for mapping a time series, x_i, to another time series of multidimensional feature vectors: $f(x_i) : R^{n \times m_i} \mapsto R^{k \times l_i}$ Deep learning has recently disrupted this process and substituted the whole learning in a homogeneous architecture [33]. However, regardless of what processing method is used, pre-processing is commonly seen in different learning methods, and hence is worth addressing in this chapter before introducing learning strategies. This chapter serves as an introductory to show how the pre-processing is formulated. This is of especial importance when a method is formulated for a specific practical situation. In many cases, a multidimensional time series of feature vectors carries a large amount of information as well as redundancy, and needs to be mapped to another time series constituted of feature vectors with lower dimensions without losing the important information. Sometimes the squeezed feature vectors are regarded as the patterns. Figure 3.1 shows a typical block diagram of a learning process for time series in the conventional form where a level pre-processing is invoked.

The step after the pre-processing is another level of processing towards recognizing the patterns. It is important for a classification method to preserve discriminating information both at the pre-processing, and at the recognition levels, where the former deals more with the structural contents, and the later does so merely with the temporal contents of the time series. This chapter is dedicated for describing as well as mathematically formulating the pre-processing phase needed for a learning process. The methods are partly supervised, and mainly unsupervised, but with notable considerations to im-

Feature Extraction	
Mapping to another time series	Length of the resulting time series is shorter

Pre-Processing	
Turning the features into patterns	Length of the resulting time series is not changes

Classification	
Discriminates the classes	Results in an single vector or data

Post-Processing	
Conditioning the classification output	Results in an integer number denoting the class

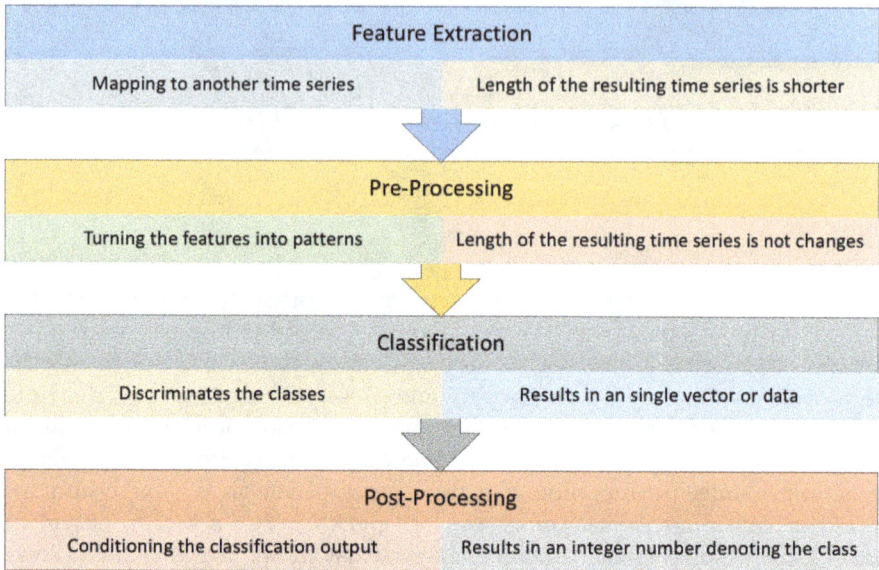

FIGURE 3.1: Classification of time series invokes these steps in its traditional form.

prove the learning process. The mathematical descriptions not only provide bases for the rest of the book, but also help the developers to build their own methods.

3.1 Dimension Reduction

In many learning problems, extracting indicative features which is capable of conveying similarity and dissimilarity between classes, leads to a large number of the features that might put the learning process in a position of high structural and empirical risks. A large feature set can result in an unstable classification in the testing phase, even though the training phase would be quick with a negligible training error. This is widely accepted that large feature vectors should be avoided, especially in a medium or small training data size, since the classifier might turn out to learn noise rather than the information from the training data. The risk of overfitting is also increased by large feature vectors.

A rule of thumb says in these cases: the number of features should not exceed more than one tenth of the training data size. The important question is: which of the features need to be omitted.

It is worth noting that some of the features might be inefficient when considered independently, but powerful when combined with other ones due to their dependent effect. Dimension reduction is indeed a mapping performed by the function $f(x_i) : R^n \mapsto R^k$ where n and k is the initial and desired number of the features, respectively. There are basically two different scenarios for this mapping: finding the mapping function f based on the statistical methods, and doing so using mathematical methods. No matter which scenario is employed, the objective is to provide powerful tools for dimension reduction. The tools in many cases are however, employed by an iterative procedure. Although there are different pursuing method for this purpose, we introduce a well-know and efficient algorithm, named "hill-climbing", for finding an optimal set of the features with the reduced dimension.

3.1.1 Feature Selection

With these methods, dimensionality is reduced based on selecting the most effective features, providing optimal discrimination power. The way of formulating the discrimination power depends on the underlying method. Here, statistical parameters of the features are estimated using the training data, and the features are selected based the estimated parameters. The two most-used methods are described in the following sections; however, different heuristic methods have been reported in the recent publications, that might be suitable for certain applications. Applicability of these heuristic methods depends on the case study. The following methods, as described by linear discrimination analysis and the Fisher method, are featured by the objectivity in their theoretical foundations that made them widely accepted. Before describing these methods, the hill-climbing algorithm is introduced, by which the two methods can be invoked for ranking and thereby reducing dimension of the feature vectors. Nevertheless, other algorithms such as genetic algorithm can be employed instead of hill-climbing algorithm, where advantages and disadvantages are points of discussions. In fact, selecting an appropriate scenario to achieve the optimal discrimination power depends on the size of the training data as well as dimension of the feature vectors.

3.1.1.1 Hill-Climbing Algorithm

This is a well-known iterative algorithm. The number of iteration equals to the number of the ultimate features desired during the feature reduction. The algorithm works based on optimization of a discriminating function used as the cost function [122][40]. At each iteration, the discriminating function is calculated, and the feature set resulting in the optimal discrimination is selected. Each input feature vector, $X_i = [x_{i,1}, ..., x_{i,n}]^T$ $(i = 1, ..., N_L)$, to the algorithm has an initial dimension of n, and the objective is to find the indices $I = [I_1, ..., I_k]^T$ $(k < n)$, whose feature set provides the optimal discrimination power. It is assumed that the discrimination power is quantified by $\mathcal{I}_{DP}$. Algorithm 1 illustrates a procedure to this end:

Algorithm 1 Hill-Climbing algorithm for finding indices of the features with optimal discrimination power

---

1: **procedure** HILLCLIMB($\langle X_i \rangle, k, \mathcal{I}_{DP}(\langle . \rangle)$)
2:     $I_1 \leftarrow \arg\max_m \mathcal{I}_{DP}(\langle X_{i,m} \rangle)$ ▷ $m = 1, ..., n$
3:     **For j=2:k**
4:         $I_j \leftarrow \arg\max_m \mathcal{I}_{DP}(\langle [X_{i,I_1}, ..., X_{i,I_{j-1}}, X_{i,m}] \rangle)$ ▷ $m = 1, ..., n$
5:     **end for**
6:     **return** $[I_1, ..., I_k]$
7: **end procedure**

The algorithm receives a feature matrix, the number of the desired dimensions, and the discrimination function as its input argument. It starts with one feature, makes an attempt to pursue for all the single features of the training data, to find the one which results in an optimal discrimination power as formulated by $\mathcal{I}_{DP}$. The index of this feature is stored in I_i. This index is then, excluded from the rest of the indices and similar pursuit is repeated, but using the previous features in the array. The discrimination function is calculated for all the new features which are independently added to the previous features, and the one with the optimal discrimination power is selected. Again, this feature is excluded from the pursue dictionary, and similar procedure is repeated with another feature added to the previous ones, and repeated totally k times to extract k features of the M ones ($k < M$). In fact, the algorithm starts from one feature and add features one by one recursively and calculate the discrimination power at the end of each iteration, and the feature set providing superior discrimination power is selected. Dimensions of the feature vector is incremented by each iteration, and indices of the features with optimal discrimination power is stored at the end of each iteration.

3.1.1.2   Linear Discriminant Analysis (LDA)

It should be mentioned from the beginning that an important presumption to this method is that all the features at each class are normally distributed. This presumption, although is not always met, but can be seen to some extent, especially when the class size is sufficiently large ($N_c > 100$) [22]. In this section, LDA is described for a problem of two classes, to simply put the reader's mind into the context, and will be generalized for multi-class problems. It is assumed that the conditional probabilities for the two classes resembles normal distribution with the mean μ_i and the covariance matrix of Σ_i ($i \in 1, 2$). Obviously, a set of the feature vectors that offers an optimal segregation between the two classes provides higher probability to be classified as class 1 when the feature vector is picked from the same class, and does so oppositely for the other class.

$$p(X_i|Q_i = 1) > p(X_i|Q_i = 2) \tag{3.1}$$

where p is the probability function for a sample from the class 1, defined by vector X_i, and Q_i is the actual class of the sample. Assuming Gaussian distribution of the feature vectors for the two classes, we have:

$$\frac{1}{|\Sigma_1|} \exp\{(X_i-\mu_1)^T\Sigma_1^{-1}(X_i-\mu_1)\} > \frac{1}{|\Sigma_2|} \exp\{(X_i-\mu_2)^T\Sigma_2^{-1}(X_i-\mu_2)\} \quad (3.2)$$

where $|.|$ denotes operator of determinant for an input matrix. In principal, without losing generality of the method, it is assumed that the probability of an input vector X_i to be belonged to the class 1 is higher than the probability of belonging to the class 2 with a certain threshold T. Applying logarithmic operations from both sides of the inequality, the following derivation is obtained:

$$\ln|\Sigma_1|+(X_i-\mu_1)^T\Sigma_1^{-1}(X_i-\mu_1)-\ln|\Sigma_2|-(X_i-\mu_2)^T\Sigma_2^{-1}(X_i-\mu_2) > T \quad (3.3)$$

Another theoretical assumption which is considered by LDA method, is homoscedasticity of the data, meaning that the covariance matrix is identical for the two classes:

$$\Sigma_1 = \Sigma_2 = \Sigma \quad (3.4)$$

where Σ is the the covariance matrix of the whole population, resulted from pooling the two classes together. Further simplification of Eq. 3.3 yields the following derivations as the decision criteria for this method:

$$\begin{aligned} W.X_i &> C \\ W &= \Sigma^{-1}(\mu_2 - \mu_1) \\ C &= W.\frac{\mu_2 - \mu_1}{2} \end{aligned} \quad (3.5)$$

The above decision criteria is employed to verify if an input vector X_i belongs to the class 1. The main assumptions for LDA are:

- Multivariate normality: Independent variables are normal

- Homoscedasticity: Covariance matrix is similar for different classes and the pooled class

- Multicollinearity: Predictive power can decrease with an increased correlation between predictor variables

- Independence: The data is randomly sampled

This is the classical form of LDA and several variants which were later introduced [145][8], however, all of these are nowadays used for either pre-processing, or for learning at the deep level [56][50][140]. In the former case, one can derive different strategies for the feature reduction process. One strategy can be based on excluding one class as the monitoring class, versus others and calculating the above threshold. Then, repeat the same procedure C times (the total number of the classes) such that each class is used only once as the monitoring class.

The threshold can be used as the discrimination power, and therefore, the set of the feature vectors maximizing sum of the thresholds can be regarded as the optimal feature vectors.

This method can be employed by a suitable searching algorithm, such as hill-climbing algorithm, for finding an optimal set of the features with the reduced dimension as you will see in upcoming sections. The reason of naming this method as linear discriminant analysis, is that the method is based on the statistical assumptions and leads to the discriminating boarders for the classes are found by which a linear transformation in Eq. 3.5, which serves as the discrimination functions. Sometime the function is found by a linear superposition of several discriminating terms. LDA was initially proposed for the classification problems, however, after development of powerful and robust methods such as deep learning, it became discoloured for such questions and mainly used for the dimension reduction.

3.1.1.3   Fisher Method

Fisher method is considered as a strong alternative to LDA, which disregards the statistical presumptions introduced by LDA, hence the applicability of the Fisher method is by far more acceptable. Nevertheless, some people categorize the Fisher method as a method of LDA, even though attentions must be paid in such a categorization. One of the point of difference is the statistical presumptions on LDA, that is not necessitated for Fisher method. Moreover, calculations for the Fisher method, although in some aspect is similar to, but generally is different from LDA, and this difference attributes especial traits to the method that will be addressed in this section. Looking back at the LDA, one can easily see that in LDA, class segregation is the main focus of the method, and threshold selection or in another term optimization, is performed based on the class differences, or alternatively the between-class variance. Here, one can think about this question, what could be the effect of feature distribution on "within classes" and also on the between-class variance [10][22]. To put this point into a better perspective, imagine a question of a case with two classes, where the data distribution is plotted in Figure 3.2.

The graph shows a case where the "between-class variance" is considerably higher than the case with lower between-class, but better situation in terms of the within-class variance. Clearly, the second case with the better within-class variance is preferred over the other case, as it can secure a superior classification, even if the data size is augmented. This condition is not foreseen in LDA with taking the within-class variance, or alternatively within-scatter, into account only. Fisher method, in contrast, relies on optimizing both the conditions, by introducing a criterion defined as the ratio of "between-scatter" to "within-scatter", as follows:

Case 1

Case 2

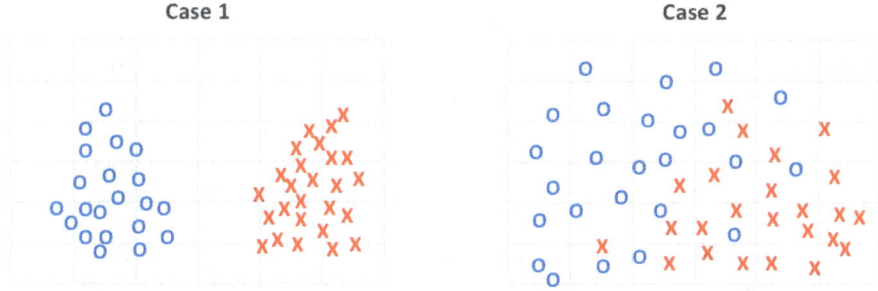

FIGURE 3.2: Data Distribution for two cases of a two class condition: low within-class and high between-class variances (Case 1), and high within-class and low between-class variations (Case 2).

$$Fisher \ \ Value = \frac{Between \ \ Scatter}{Within \ \ Scatter}$$

$$S_B = \sum_{i=1}^{C} p(\omega_i)(\mu_i - \mu) \cdot (\mu_i - \mu)^T$$

$$S_W = \sum_{i=1}^{C} p(\omega_i) \cdot \Sigma_i$$

$$\Delta = \frac{|S_B|}{|S_W|} = \frac{\left| \sum_{i=1}^{C} p(\omega_i)(\mu_i - \mu) \cdot (\mu_i - \mu)^T \right|}{\left| \sum_{i=1}^{C} p(\omega_i) \cdot \Sigma_i \right|}$$

$$\mu_i = E\{x|\omega_i\}, \quad \mu = E\{x\} = \sum_{i=1}^{C} p(\omega_i) \cdot \mu_i$$

$$\Sigma_i = E\{(x - \mu_i) \cdot (x - \mu_i)^T |\omega_i\}$$

(3.6)

where $p(\omega_i)$ is the probability density function of the random variable ω_i. Δ is called Fisher value which is in fact regarded as the discrimination function for this method [50]. The Fisher value is simplified to the following derivation for single valued numbers rather than the vectors:

$$\Delta = \frac{(\mu_1 - \mu_2)^2}{\sigma_1^2 + \sigma_2^2} \tag{3.7}$$

where σ^2 is the variance value of the feature. If the within-scatter approaches a singularity condition (determinant of the within-scatter becomes zero), the Fisher value tends to infinity, and hence the discrimination power comparison cannot be performed in a realistic manner. This condition happens when the feature vectors lose their independence in terms of the linear algebra. If at least two dependent vectors exist in S_W, the two columns corresponding to the dependent vectors are aligned in the same orientation as they are linearly dependent. In this case, the determinant of S_W, becomes zero. This cannot often happen unless feature repetition is avoided, since there is always a level of noise, associated with the data, causing differences even for the dependent vectors, and securing a non-singular matrix of the within-class variance.

Fisher method can be independently employed for individual features one by one, to rank the features according to the Fisher value, and select the ones with the highest ranks. A certain number of the features with minimal Fisher value can be discarded. A high Fisher value corresponds to a well segregation, or alternatively high discrimination power. Fisher method can be invoked by the hill-climbing algorithm (described in the previous sequel), in order to keep feature dependence regarded in the discrimination. In some of the applications, a single feature is not as effective as when it is considered in conjunction with other features. This justifies the importance of considering feature dependence in the learning process, which lays well into the topic of multi-variate analysis.

3.1.2   Linear Transformation

In a group of the methods for feature reduction, a matrix of $\Re^{k \times n}$ $(k < n)$ is found, which serves as the linear transformer to reduce dimension of an input feature vector (n and k are the initial and ultimate dimension of the feature vector, respectively). Principal component analysis and factor analysis are two well-known methods to this end, broadly used in many applications, not limited, but well-beyond to feature reduction. Nowadays, principal component analysis is becoming popular not only as an efficient mathematical tool for feature reduction, but also as an important part of many classification methods and learning systems. It is therefore, essential to address these methods especially for the beginners. One should keep in mind that both of the following methods are categorised within the context of statistical methods, where statistical distribution of features plays an important role in the implementation [7]. The main point of difference separating them from LDA lays in the feature transformation for these two, in contrast with LDA in which a process of feature selection is pursued.

3.1.2.1   Principal Component Analysis (PCA)

Let us assume that the training feature set is arranged in a matrix W of $n \times N$, where n is the dimension of the feature vectors and N is size of the learning data (for simplicity in writings, N_L is replaced by N). PCA does not discard any feature, instead, applies a linear transformation to the feature set, which is interpreted as feature projection on k orthogonal vectors. The resulting vector is obtained from the rotation of the feature vectors (k is the desired number of the feature vector) in the vector space. In PCA, the rotation is guaranteed to be done in a way to provide the maximum dispersion along each direction. The orientations of the rotated vectors are mutually orthogonal to all other vectors, and the projection of a feature vector on the unity vectors that shows the direction, is known as the principal component of the feature vector on that direction. Figure 3.3 depicts a two dimensional case. The rotation to the new directions is shown in the figure.

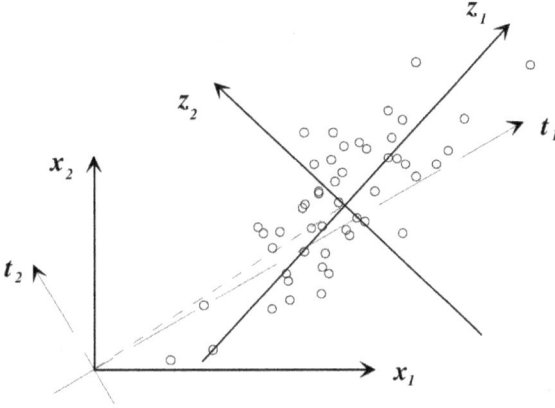

FIGURE 3.3: Principal component analysis rotates the data, or in another view the axis, in a way to provide maximum dispersion along the rotated axis.

As can be seen in the figure, PCA is an effective method, when the data dispersion is not at the same direction with respect the coordinates. PCA implementation involves two steps:

1- finding the covariance matrix of the feature matrix $W = [X_1, ..., X_N], X_k \in \mathbb{R}^n, k = 1, ..., N$

$$\mathcal{K}_{W,W} = \begin{bmatrix} E_{W_1,W_1}, ..., E_{W_1,W_N} \\ ... \\ E_{W_N,W_1}, ..., E_{W_N,W_N} \end{bmatrix} \tag{3.8}$$

2- Then, the eigenvectors of the covariance matrix, $\mathcal{K}_{W,W}$, are calculated and ordered according to the eigenvalues in the descending order. The first k ($k < n$) eigenvectors are employed to find the principal components, defined as the projection of the feature vectors on those eigenvectors.

It is important to note that normalizing dynamic range of the feature vectors before performing the rotation, can substantially improve capability of PCA in feature reduction. Algorithm 2 shows a pseudo algorithm for the PCA implementation, written in compliance with the MATLAB® codes:

Algorithm 2 receives a matrix of feature vectors W ($W \in R^{n \times N}$), in which each feature vector has a dimension of n, as well as the dimension of the feature vectors k ($k < n$). The feature matrix is then linearly transformed, and the transformed features with the reduced dimension is returned by the the procedure in Z. Normalization is regarded as an important step of the process. The richer dataset with a broad coverage of data is, the better performance with reduced dimension it results.

Algorithm 2 Standard Principle Component Analysis

1: **procedure** PCA(W, k)
2:   ▷ Beginning of normalization
3:     $M \leftarrow [\frac{\sum_{i=1}^{N} W_{1,i}}{N}, ..., \frac{\sum_{i=1}^{N} W_{n,i}}{N}]^T$ ▷ Average Calculation
4:     $S \leftarrow [\sqrt{\frac{\sum_{i=1}^{N}(W_{1,i}-M_1)^2}{N-1}}, ..., \sqrt{\frac{\sum_{i=1}^{N}(W_{n,i}-M_n^2}{N-1}}]^T$ ▷ Standard Deviation
5:     **For j=1:N**
6:         **For i=1:n**
7:             $Y(i,j) = \frac{(W_{i,j}-M_i)}{\sqrt{N}S_i}$
8:         **end for**
9:     **end for**   ▷ End of normalization
10:    $K = Y * Y^T$ ▷ Covariance of the normalized features
11:    $[V, \lambda] = eig(K)$ ▷ Eigenvector V, and eigenvalue λ
12:    **For j=1:k**
13:        **For i=1:n**
14:            $R = \frac{V_{i,k}}{\sqrt{\lambda_k}}$ ▷ Transformation matrix
15:        **end for**
16:    **end for**
17:    $Z = R^T * Y$ ▷ Transformed features
18:    **return** Z
19: **end procedure**

In some of the applications, it is required to perform a feature reduction for the purpose of removing the background noise associated with the data. In this case the number of the desired dimension k is not known, and should be found automatically, where the process must be continued until to remove a certain noise level. In such cases a certain threshold is empirically assigned, and the eigenvectors with the eigenvalues of less than the assigned threshold will be suppressed. Keeping the eigenvectors above the threshold can provide a mean for finding an optimal value for k.

Even though, PCA is widely employed in different applications, it is not an absolute drawback-free method. A criticism which is associated with PCA lays in losing interpretablity of the resulting features with the reduced dimension. In fact, interpretability of the features is lost after rotation where the transformed feature vectors will have no physical meaning anymore, whereas other methods of feature selection, in which the feature vectors with the reduced dimension still have their initial meanings.

3.1.2.2   PCA-Fisher Method

It was shown in the previous section that the main art of PCA is in the rotation of the coordination in the feature domain in a way to provide dispersion along the access. The eigenvector corresponded to the highest eigenvalue points to the direction of the coordination with the highest dispersion. In most of the cases, after the first few eigenvalues, this value is substantially dropped, implying on the fact that the contents of the features do not convey much information, or typically contain noise. As a consequence, PCA has the capability to be used in an unsupervised manner for feature reduction. To this end, the trend of the decreasing eigenvalues is investigated and the one corresponding to a significant decline in eigen value would be considered as the reduced size of the feature vectors. Similar trend can be considered, when the Fisher method is employed for feature selection. The main difference, between the PCA and Fisher method resides at this point that PCA transforms an input feature vector to another one with lower dimensions and different feature values, however, in Fisher method, only dimension is reduced and the feature values remained unchanged. An important question raised here at this point is: which method is preferred as the most efficient one, if we are obliged to preserve optimal data dispersion? To answer this question, you need to consider that Fisher value and PCA value, are used as the indicative numbers showing discrimination and dispersion, respectively. Figure 3.4 depicts different cases in terms of the accordance between discrimination and dispersion.

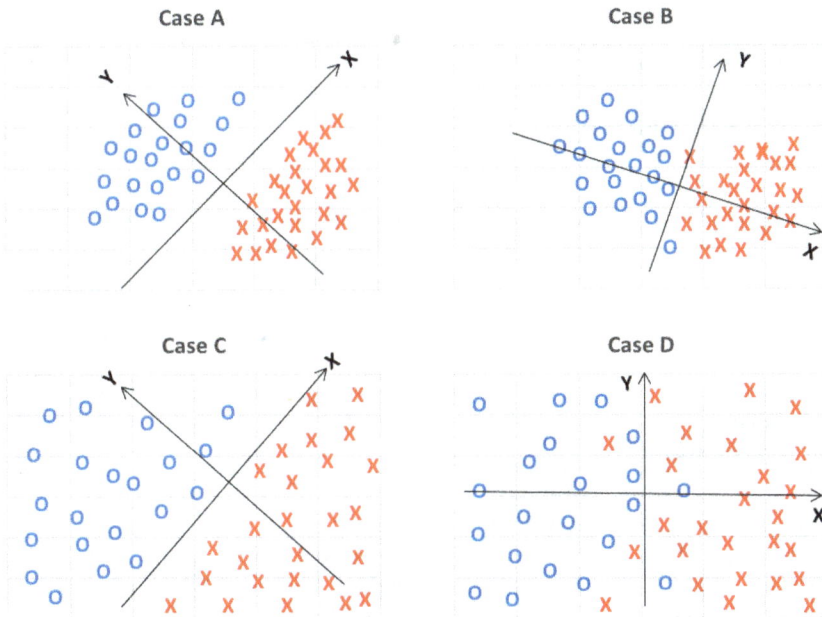

FIGURE 3.4: Comparison of principal component analysis to the Fisher discriminant analysis for a classification problem. In Case A, both the methods can result in a good discrimination power. In Case B, Fisher method can provide a better discrimination. Oppositely, in Case C, principal component analysis can result in a better discrimination power. In Case D, both the methods are impotent to provide a good discrimination.

Figure 3.4A illustrates a case of two classes of 2-dimensional feature vectors. In this case, dispersion and discrimination lay along the same direction.

With this distribution of data, a large dispersion along the rotated x axis is observed, which is in correlation with the discrimination, and hence both the PCA and the Fisher method, equivalently imply that the feature in the rotated y axis gives the highest discrimination power. Figure 3.4B, shows a case in which dispersion is not in accordance with the discrimination. It is clearly seen that discrimination is better provided by the Fisher value, while PCA cannot deliver an efficient reduction. This is in contradiction with Figure 3.4C where PCA can serve as a more powerful method than Fisher in feature reduction, and lastly Figure 3.4D demonstrates a case in which both the methods cannot deliver inappropriate performance. In such the case, a combination of the two method is recommended. Here, the PCA method followed by the Fisher method will substantially improve the performance, since a rotation results in an improved dispersion, and the Fisher methods guarantees a high value of the discrimination power. One way to this end is the use of PCA method for finding the transformed feature matrix, alternatively saying the rotated feature matrix, followed by the Fisher method for selecting those ones.

Algorithm 3 Combined PCA and Fisher method through the Hill-Climbing algorithm

---

1: **procedure** PCA-FISHER($\langle X_i \rangle, k$)
2:     $W \leftarrow [X_1, ..., X_N]$ ▷ Feature matrix $W \in R^{n \times N}$
3:     $Z \leftarrow PCA(W, n)$ ▷ PCA transformed matrix
4:     $\mathcal{I} \leftarrow HILLCLIMB(Z, k, \Delta)$ ▷ Δ is the Fisher discrimination function defined in (3.6)
5:     **For j=1:k**
6:         **For i=1:N**
7:             $\mathcal{Z}_{j,i} = Z_{\mathcal{I}_j,i}$ ▷ Transformation matrix
8:         **end for**
9:     **end for**
10:     **return** $\mathcal{Z}$
11: **end procedure**

This combination improves cases with heterogeneous dispersion [10]. It is worth noting that PCA method or any kind of the related methods can offer a more suitable performance only when contents of the information is far higher than the noise at the feature level. This should be also taken into consideration for the time series analysis. Obviously, absence of this condition may mislead the feature reduction process by **performing an improper rotation towards dispersion of the noise contents.**

3.2   Supervised Mapping

A single multidimensional data sometimes carries unnecessary information, which can mislead any classification task. Noise is a common source of such redundancy, and always superimposes on the data, and occasionally overwhelms the information, such that the feature contents are heavily contaminated and it is not easy to extract the pure information from the features. In most of the classification questions, there is often an intention to decrease redundancy in multidimensional data by pruning additional contents. Apart from the statistical methods, described in the previous section, one way is to map the multidimensional data (usually high dimensional), to the categorical forms, or to the numerical symbols, in a pre-processing phase. The process is sometimes called "pattern detection", which is often followed by another phase named "pattern recognition", in a multi-step learning process. Certainly, this structure is not followed in all the learning scenarios, but privileged especially in hybrid learning methods [11][151][31]. In this continuation, basics of neural networks, as the well-known methods, widely accepted by the artificial intelligence community, will be described in detail.

The idea of the neural network has been inspired by activities in human brain cells, and has tried to imitate the learning process happens in the human's brain [137][81]. It is obvious that complexities in the human brain attributes a high level of approximation with such imitations, however, even simplified implementations have been responding to many of the practical questions. Single layer perceptron, as the building block of any neural network architecture, as well as multi-layer perceptron are introduced in this section as the efficient methods for supervised learning. Before diving into the theory of neural network, a well-known supervised method, named K-Nearest Neighbours (KNN), is briefly described. Although popularity as well as versatility of KNN couldn't reach neural networks, it is worth to mention KNN as it is widely employed by AI community as a part of the classification system and even in a number of the hybrid structure [133]. Regardless of the classification method, we make the following assumptions, which will be commonly used for all the supervised methods.

There is a training dataset, V, with the priory known classes and the following parameters:

$$V = \{(X_i, Q_i) : X_i \in R^n, \ Q_i \in \{1, ..., C\}, \ i = 1, ..., N\} \qquad (3.9)$$

- N : Size of the training data

- C: Number of the classes

- X_i: Input vector

- n: Dimension of the feature vector

- Q_i: Actual class of an input

3.2.1 K-Nearest Neighbours (KNN)

KNN is a supervised classification method. The K nearest neighbours of a testing sample, Y, $(Y \in R^n)$ are found using Euclidean distance:

$$D_i = (Y - X_i)^T * (Y - X_i) \tag{3.10}$$

Then, the class which receives majority of the votes in the vicinity of K samples around the testing sample, will be assigned as the sample class. Algorithm 4 describes the KNN method:

Algorithm 4 Classification of a testing sample, Y, based on KNN method

---

1: **procedure** KNN($\{V\}, Y, K$)
2:    Calculate $\{D_i(Y, \{X_i\})\}$ ▷ Using Eq. (3.10)
3:    $S = Sort(\{D_i\}, Descending)$
4:   ▷ Sort the distances in the descending order
5:    $S_T = \{S_i : i = 1, ..., K\}$
6:    $I \leftarrow \arg\{S_T\}$
7:    $q \leftarrow \arg\max_i\{Q_i(I)\}$ ▷ q is the classification result for the sample Y
8:    **return** q
9: **end procedure**

3.2.2 Perceptron Neural Network

Perceptron neural network is a basic building block for linear classification which has been inspired by the action of neural cells of human brain, with similar hypothetical function. Although neural function of a human brain has not been profoundly understood yet, but several studies attempted to model parts of the neural functionality such as learning process [137][153]. The extent to which the models exceed depends on the underlying function in conjunction with the corresponding information extracted from the neurological behaviours. Based on these findings, a simplified model of neurological activities, in sense of learning process, has been proposed, that received much interests from the researchers such that many people recognised it as an initiation of artificial intelligence [97]. A single neuron of human brain, as a basic building element of the neural system, is composed of a nucleus, dendrites, axon and synapses. Figure 3.5 demonstrates this model.

An introductory description to neural activities is worth mentioning at this point. In this model, the inputs from other surrounding neurons are received by the dendrites, and sent to the nucleus of the neuron, which acts as a corpus. Depending on the inputs as well as the ionic condition of the nucleus, an output is created by the nucleus, and delivered to the axon connected to it. The axon transfers the message received from the nucleus and passes it to several synapses to perform a neural action. This neurological model, which is fully justified by the physiological bases, is mathematically modelled by a number of the elements like corpus, inputs, outputs and weights, simulates

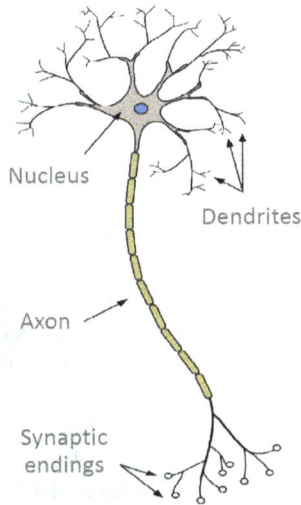

FIGURE 3.5: Neural simulation model used as the building block of neural networks.

the action of nucleus, dendrites, axon and synapses, respectively. This model is named a perceptron, introduced in 1940th by McCulloch and Pitts, which is demonstrated in Figure 3.6.

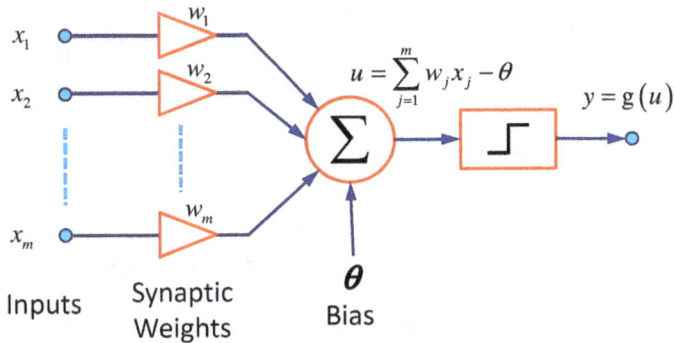

FIGURE 3.6: A single neuron perceptron model for simulating neural function.

In this mode, a neuron receives a superposition of its inputs (sum of the weighted inputs) in conjunction with a bias value that resembles the neural activity. If the sum of the weighted inputs exceeds a certain threshold (neuron bias), the output of that neuron becomes activated. The output y of a neuron, for an input vector $X = [x_1, ..., x_m]^T$, in the perceptron model is calculated as follows:

$$y(X_i) = g(W^T X_i - \theta) \tag{3.11}$$

In this model, g is the ACTIVATION FUNCTION, W is the weighting vector, and θ is the bias value. The activation function in the linear model is the Sign function:

$$g(x) = Sign(x) = \begin{cases} 1 & if \quad x > 0 \\ 0 & otherwise \end{cases} \tag{3.12}$$

The neuron's output $y(X)$ for an input X becomes 1, or alternatively saying "is activated", when the weighted superposition of the inputs $W^T X$, added by the bias value θ, becomes positive (> 0), otherwise the neuron remains inactive. This is in analogy to the humans' neural cells; the neuron Axon becomes activated (an action potential is created) when the superposition of the connected dendrites exceeds a certain threshold of the ionic quiescence. This model in its nature, performs a binary classification of two states: activated inactive. In a binary case, if the model parameters, W and θ, are appropriately selected, the neuron's output y gives identical state for majority of the vectors from the same class.

For multi-class problems, one neuron is independently assigned to each class, whose output must become 1 only for that specific class, and zero for the rest of the classes, as depicted in Figure 3.7.

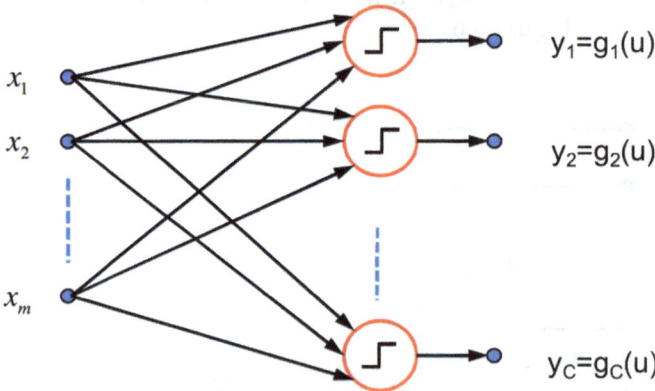

FIGURE 3.7: Multi-class representation of perceptron model, named perceptron neural network.

In this case, the weighting parameter is constituted of a set of the weighting vectors, each assigned for a neuron, and altogether constituting the weighting matrix:

$$W = [W_1, ..., W_C]^T \tag{3.13}$$

Assigning a separate bias θ_j to each neuron, the outputs becomes:

$$y_{i,j} = g(W_j^T X_i - \theta_j)$$
$$j = 1, ..., C \tag{3.14}$$

The values of $\{y_{i,j}\}$ is the model output, resulted from the perceptron model, or in other words, the classes obtained by the model. The actual class, $\{Q_{i,j}\}$ ($\forall i,j \; Q_{i,j} \in \{0,1\}$) for each data is composed of a vector of the length C, in which, only and only, one element which corresponds to the related class equals to 1, and the rest of the elements are zero:

$$\forall j \; \exists ! \; q_{i,j} : q_{i,j} = 1$$

$$\forall i : \sum_{j=1}^{C} q_{i,j} = 1 \tag{3.15}$$

If a reckoning perceptron neural network is perfectly trained (which is an absolute definition), the same rule implies on the output all the neurons of the neural network. However, in the practical cases, this ideal situation scarcely happens with large data size, and some output neurons other than the actual class of the output vector become 1.

$$\exists i : \sum_{j=1}^{C} y_{i,j} \neq 1 \tag{3.16}$$

Training of a perceptron model, leads to the optimal values for W and θ, to provide an optimal classification performance. It is performed in a supervised manner. Likewise in all training methods, a criterion must be defined to quantify the classification performance. In the perceptron model, the least square error, $\mathcal{I}$, calculated based on the subtraction of the predicted class and the actual class for each input, is invoked as the criterion for the training. In order to preserve both the negative and positive errors, sum of the least square error is employed as the criterion for the training:

$$\mathcal{I} = \sum_{i=1}^{N} \sum_{j=1}^{C} \left(y_{i,j}(X_i) - q_{i,j} \right)^2 = \sum_{i=1}^{N} \sum_{j=1}^{C} \left(W_j^T(X_i) - \theta_j - q_{i,j} \right)^2 \tag{3.17}$$

$$i = 1, ..., N$$

The bias values can be easily incorporated into the weighting matrix. Although an optimal result can be found by taking the derivation from (3.17), iterative procedures are always preferred in order not to encounter with singularity. This will yield an updating procedure for the W which incorporates θ as well:

$$\frac{\partial \mathcal{I}}{\partial W_j} = 0$$

$$W_j = (X * X^T)^{-1} \mu Q \tag{3.18}$$

where Q and μ ($0 < \mu < 1$) is the actual class and the learning rate, respectively. However, taking the inverse of matrix might lead to singularity. In practice the learning parameters are iterative found. The values are randomly initialized first and then updated through the following recursive formulation:

$$W_j^{t+1} = W_j^t + \mu(q_{i,j} - y_{i,j}^t) \tag{3.19}$$

where t denotes the iteration number, so called epoch. The bias θ is also treated as a weight in W with the consistent input of 1. The The training is ultimately stopped at reaching certain criterion, typically reaching a certain number of epochs, or a certain low level of the error defined by Eq. (3.17), or even a combination of the both. Perceptron model provides a linear classification, which can be insufficient for many practical cases, that are typically more complicated.

Perceptron neural network offers "DISCRIMINANT LEARNING" which is regarded as a positive aspect of this method. The term "discriminant learning" is used for those methods in which learning one class of the data during the training phase, does not affect learning the parameter of the other classes. In the perceptron model for example, the learning weight vectors W_j are independent for each class, without having any shared weight. As a result, training neuron j, assigned to a certain class, cannot put any influence on the learning weights of the other neurons.

3.2.3 Multi-layer Perceptron Neural Networks (MLP)

An arrangement of the perceptron neural network, in several sets of the neurons, whose outputs constitute the inputs of other neurons, is defined as Multi-Layer Perceptron neural network (MLP). MLP performs a nonlinear mapping from R^n to $\{0,1\}^C$, therefore offers a fine border between the classes with the capability of learning complex patterns comparing to the linear ones, i.e., Perceptron model. Figure 3.8 illustrates MLP architecture.

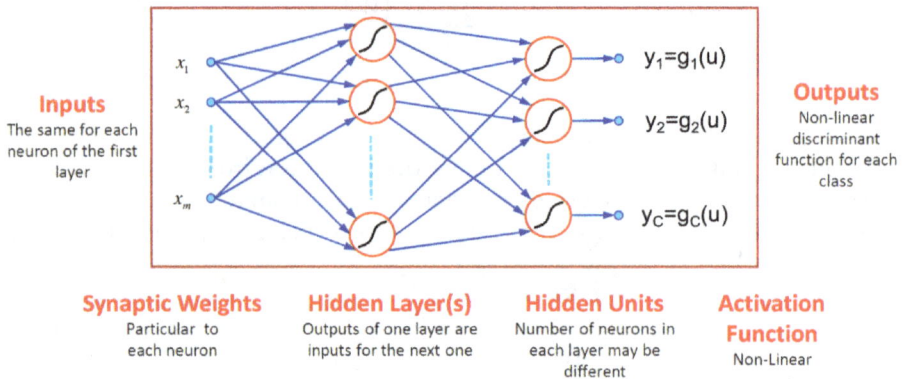

FIGURE 3.8: Multi-class representation of perceptron model, named perceptron neural network.

The input of a MLP is named, the input node, the middle layer(s) is(are) named the hidden layer, and the set of the neurons whose outputs perform the classification, is named the output layer. The number of the neurons in the output layer depends on the number of the classes in the training data. Output of MLP is derived by the following formula:

$$\phi_j(X_i) = g_2(W_{2,j}^T \cdot g_1(W_{1,k}^T X_i - \theta_k) - \theta_j) \tag{3.20}$$

Unlike a single layer perceptron where the classification method was linear, the neurons' activation function of MLP is a nonlinear function, e.g., Sigmoid, which will be described in the followings. Neurons' activation function of a MLP must follow certain conditions in order to guarantee the learning process, termed by convergence.

Activation function of the neurons in the MLP structure must be continuous, derivable, and limited between -1 and 1. Logistic Sigmoid, is one of the most common activation functions, which is formulated in (3.21) and depicted in Figure 3.9.

$$g(u) = \frac{1}{1 + \exp(-\beta u)} \tag{3.21}$$

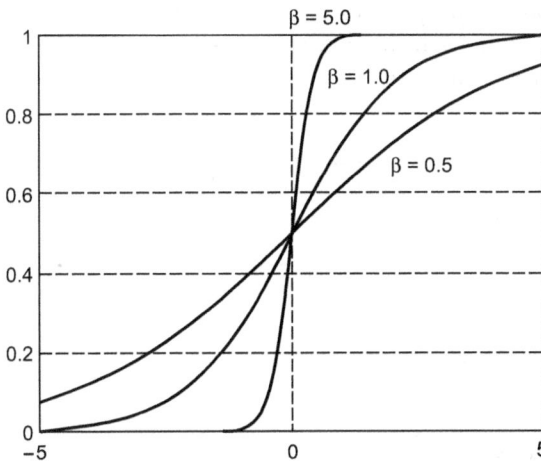

FIGURE 3.9: Graph of sigmoiid function for different values of β, which is used as the activation function for the perceptron models.

The parameter β affects the sharpness of the neurons' activation, as seen in the figure, but has not very much influence on the training process such that the training remains approximately unchanged for a broad range of β.

Training of a MLP is based on calculating least square error as the learning rule (see Eq. 3.17). The training is performed in a supervised manner, through a specific and well-known method, named BACK-PROPAGATION

ERROR method. As with the single layer network, the training method is an iterative process for MLP, where the least squat error calculated from the distance between the actual class and the ones obtained by the method, that is attempted to be minimized at each iteration.

$$\mathcal{I} = \sum_{i=1}^{N}\sum_{j=1}^{C}(\phi_{i,j}(X_i) - q_{i,j})^2 = \sum_{i=1}^{N}\sum_{j=1}^{C}\left(W_j^T(X_i) - \theta_j - q_{i,j}\right)^2 \tag{3.22}$$

$$i = 1, ..., N$$

The classification is discriminant, and as with the multi-class perceptron, "1" is set for the output neurons when the input comes from the learning class, and "0" for the rest of the neurons. The weights are updated in a way to minimise gradient of the error on both the output, and the hidden layers. Considering the logistic Sigmoid activation function, derivation of the logistic Sigmoid function becomes:

$$g(u_{j,k}) = \frac{1}{1 + \exp(-u_{j,k})} = y_{j,k}$$

$$\frac{\partial g(u_{j,k})}{\partial u_{j,k}} = \frac{\exp(-u_{j,k})}{1 + \exp(-u_{j,k})} = y_{j,k}(1 - y_{j,k}) \tag{3.23}$$

The error gradient propagates from the output layer towards the input layer and the error at each layer can be easily calculated using Eq. (3.23). Figure 3.10 illustrates the error propagation:

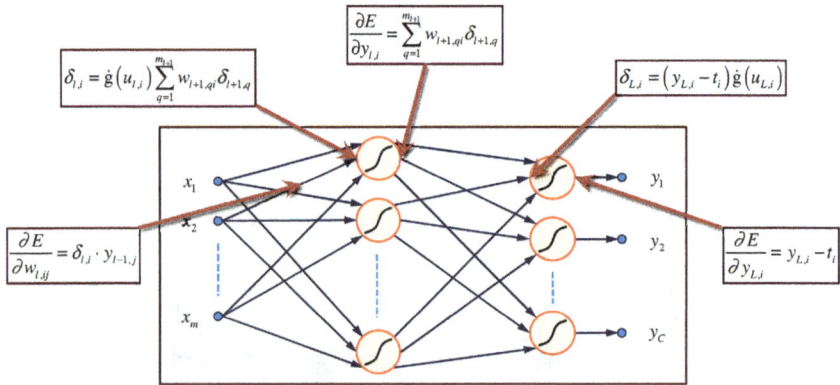

FIGURE 3.10: Propagation of the error from the output layer to the input layer of a multi-layer perceptron neural network.

The training process is based on initialization of the neurons with random values, and updating their weights by the following formula:

$$W_{i,j,k}^{t+1} = W_{i,j,k}^t + \alpha\frac{\partial \mathcal{I}}{\partial W_{i,j,k}} \tag{3.24}$$

Nowadays, powerful toolboxes in some research programming languages such as MATLAB and Python offer easy implementation, without the need to deeply digging into the learning principals. To a lesser extent, for those who are interested in understanding the calculation contents, details of finding the learning parameters through an iterative algorithm can be found in many older texts [4][81].

It is obvious if the number of the neurons and the layers tend to be high, the training process will need more time to appear its effect on the output, tending to make the process slower. It is certain that there is no closed formula for finding the number of the hidden layers and also the number of the neurons at each layer. Although there is no explicit rule to offer an appropriate architecture, it has been experimentally shown that exceeding a certain number of the neurons cannot improve the classification performance, and may increase the structural risk without gaining any advantages.

From Eq. (3.24) one can easily see that the sample number i comes to the training phase. This point opens the discussion on how to update the learning weights, for example, shall we calculate the error for one single sample and then update the learning parameters after calculating the error for each sample, or to calculate the error for all the samples of the training dataset, then update the learning wights? This leads to different schemes of the training method. Updating the weights are basically performed in either of the two fashions: batch training or online training. In the batch training the weights are updated after presentation of the whole training dataset, where cumulative gradient is employed (see Eq. 3.25), while in the online training the update is performed after presentation of each single data (Eq. 3.26)

$$\Delta W_{i,j,k}^t = -\alpha \sum_{i=1}^{N} \frac{\partial \mathcal{I}}{\partial W_{i,j,k}} \tag{3.25}$$

$$\Delta W_{i,j,k}^t = -\alpha \frac{\partial \mathcal{I}}{\partial W_{i,j,k}} \tag{3.26}$$

The coefficient α ($0 < \alpha < 1$) in the above equations is known as the learning rate, which gives a control on the learning speed. Choosing a high α near 1 makes the training process quicker, but coarser, implying on the poor classification performance. In contrary, a low α, is an indication of a slow training, but fine decision boarders between the classes. A complete process of presentation of the training data followed by updating the learning weight, is called one training "epoch". An epoch demands longer time for batch training, compared to on-line training, with similar training data. An iterative process of training is often stopped at certain criterion, typically reaching a certain number of epochs, or a certain low threshold of the classification error defined in Eq. (3.22), or a combination of the both. In general, the batch training is a rather quick learner comparing to the online training, even though it requires longer epochs. The reason is the quicker convergence in the batch training in which the error drops much quicker than the online training. However, the online training offers a lower risk to be trapped in the local minimum, a condition in which the error cannot decrease with more training epoch, while potentially the neural network has the capability of lower classification error. In such the cases, the training is usually stopped and restarted from the beginning to run around the local minimum. In any of the cases, in order to maintain the training process with lower number of the required epochs, another term, named as "momentum term" is added to the weights. The momentum term is defined as follows:

$$\Delta W^t_{i,j,k} = -\alpha \frac{\partial \mathcal{I}}{\partial W_{i,j,k}} + \beta \Delta W^{t-1}_{i,j,k} \tag{3.27}$$

This would reduce sensitivity to learning with the noise, in gradient estimation. However, implementation of the momentum requires skill in finding proper values for α and β, since very low values makes a negligible effect as the gradient, and on the other hand very large values create error oscillation. A solution for assigning a proper value for the learning rate is the use of the adaptive learning rate, which changes with the training epoch. Vogl learning algorithm is a well-known solution, in which the learning formula is given by:

$$\Delta W_{i,j,k} = -\alpha^t \frac{\partial \mathcal{I}}{\partial W_{i,j,k}} \tag{3.28}$$

and the value of the alpha varies in time according to the following rule:

If $\mathcal{I}^t < \mathcal{I}^{t-1}$: $\alpha^{t+1} = \alpha^t(1 + \epsilon_a)$

If $\mathcal{I}^{t-1} \le \mathcal{I}^t < (1 + \rho)\mathcal{I}^{t-1}$: $\alpha^{t+1} = \alpha^t$

If $\mathcal{I}^{t-1}.(1 + \rho) \le \mathcal{I}^t$: $\alpha^{t+1} = \alpha^(1 - \epsilon_d)$

This algorithm is by far less sensitive against the parameters: ρ, ϵ_a, and ϵ_d, which are all real numbers between 0 and 1.

Back to the batch and online training, in many training procedures, a combination of batch and online training is preferred, as in a well-known method named, "diagonal Levenberg-Marquadt" algorithm, part of the training data is employed to calculate elements of a matrix, named Hessian as follows:

$$\Delta W^t_{i,j,k} = -\frac{\alpha}{\dagger + H_{xx}W_{i,j,k}} \frac{\partial \mathcal{I}}{\partial W_{i,j,k}} \tag{3.29}$$

The Hessian matrix is utilized for the updating the weight in on-line manner:

$$\{H_{xx}(W_{i,j,k})\} = \{\frac{1}{L}\sum_{k=1} C \frac{\partial \mathcal{I}}{\partial W_{i,j,k}}\} \tag{3.30}$$

Therefore, both batch and on-line fashions are employed for this method. Diagonal Levenberg-Marquadt, is a very quick learner method, with relatively low number of epochs, needed for a certain value of the training error, as comparing to other alternatives. However, a larger training memory is required for the arithmetic manipulations. The readers are encouraged to read relevant textbooks for more details of this method, which is beyond the scope of this book.

MLP has been broadly employed, either by itself, or as a pre-processing method in hybrid classification for detection or recognition of patterns in a stream of time series. Simplicity in training, in conjunction with the availability of this technology, which was in turn caused by its intrinsic versatility, made MLP, a popular solution in different applications. Earlier researchers in 1980th, were sometimes considering MLP as a black box alternative to the statistical classification methods like KNN, and MLP was regarded as a blind learning method instead. The connection between statistical methods and MLP was later understood, and the unwritten confrontation, or at least reluctance, from the earlier researchers was removed when potentials of MLP were further explored [115][15]. In particular, it has received especial attention from the researchers after which its link to the statistical classification methods has been understood [115][15][14].

3.3 Unsupervised Mapping

In most of the advanced learning systems, there are many practical situations where we do not, or essentially sometimes cannot, have access to the label of the training data. In other cases, it sometimes happens that the access is not provided in the middle level of the process, even though the endpoint labels are accessible. This necessitates invoking suitable unsupervised methods for a level of pre-processing to reduce dimension of the feature vectors, either to avoid high structural risk possibly associated with the redundant features, or to prune extra features that might be created by contaminating noise. The training phase of unsupervised methods is mostly recursive, starting with a random initialization, and terminating with a stable situation of categorising the training data. This categorization of a dataset in an unsupervised way, is known as "clustering", in contrast with classification which implies on a supervised learning. The following subsections describe a number of the efficient and common methods, invoked for the purpose of clustering.

3.3.1 K-Means Clustering

Clustering by definition is known as an unsupervised assignment of a label to a dataset. The assignment is usually applied based on the similarities among the dataset, such that similar data receive identical cluster label. This is an explicit case of learning. In clustering, numerical symbols are often used for labelling the clusters.

K-Means method performs a mapping of M dimensional feature vectors of real numbers, $\{X_i : X_i \in R^M\}$, to one dimensional vectors of integer numbers, $\{q_i : \forall i, \ q_i \in \{1, ..., K\}, \ K < M, \ i = 1, ..., N\}$ (N is the training data size). Thus the number of the clusters, K, is known prior to the learning process. This number is sometimes, known as the "quantization level" and treated as hyperparameter. Remembering from the Fisher method, where distribution of the data was taken into account in sense of the both within and between classes, K-Means fundamentally does not assign any weight to the distribution, and merely deals with the discrimination. Nevertheless, in a case of a single feature with Gaussian distribution, K-means can eventually end-up with the clustering result similar to Fisher or LDA method.

K-Means clustering method is based on recursive procedures begins with an initialization and ends-up with a stable quantified label for each input data. Stability in this definition means that further iteration does not noticeably change the labels. This procedure is:

- Centroid Initialization: Initial K feature vectors at R^M by random

- Classification: Classify the data to K classes according to their Euclidean distances to the centroids

- Update: Update the centroids by the average value of the data at each class

- Termination: Continue until reaching stable values for the centroids

K-Means in its original form does not consider any assumption for initialising the recursive procedure. It assigns K data, out of the N training data ($K < N$), as the available samples, for calculating the initial values of the centres of gravity by random. These centres of gravity are named as "centroid". Each member of the training data is assigned to the nearest centroid based on calculation of the minimum of Euclidean distance. After assigning a class to all the data members, the centroids are updated by the average value of the feature vectors at the same cluster. This is the end of the one whole recursion. The same procedure is repeated several times until reaching stable centroids.

K-Means provides a very simple and relatively efficient clustering method. Although intra-cluster variance is deemed to be minimized during the learning process, the global variance is not guaranteed to be minimised by this method. Another drawback of this method is that the centroids which are close to each other, do not necessarily correspond to similar values in the feature space. It is important to note that many literatures associate another drawback with K-Means clustering method, which is random cluster assignment to the data caused by random initialization of the centroids. This may not be a real drawback. A simple solution is to introduce a certain rule for initialisation of the centroids. One can suggest assignment of the centroids according to the norm of the feature vectors. In this method, norm of the data is calculated for all the samples and uniformly assigned to the centroids according to their values. In this rule, the largest and the smallest norm of the feature vectors are assigned as the highest and the lowest borders. Then, the rest of the data is sorted according to their norms. The values between the norms are uniformly divided into K values, and regarded as the reckoning norms of centroids. The feature vectors with the nearest norms to the reckoning norms are selected as the initial centroids. It is evident that such the initialisation prevents random settings of the centroids, and therefore guarantees reaching stable clustering.

3.3.2   Self-Organizing Map (SOM)

Classification, in its functional mapping view, can be interpreted as a way of quantification or dimensional reduction, even though this aspect is not always considered by the existing definitions, and especially by the contextual taxonomy. If dimension reduction is decreased down to 2 or 3 dimensions, the processed high dimensional data can be visualized in a surface or space. That's why dimension reduction can be used as a way of visualization. A good visualization method should provide interpretative, understandable and rational distribution of data in a 2, or 3, dimensional space. In practice, a training dataset is always contaminated by different sources of the noise, causing hard segregation, questionable. The borderline data, caused by noise or any other source of irregularity, are typically expected when it comes to a practical visualisation. Many classification methods, such as MLP, employ nonlinear mapping that set a well segregation to the data, such that the borderline data is assigned either correctly, or incorrectly in one of the classification groups. This associates an uncertainty to the methods. This feature of the classification methods make them inefficient for the visualisation.

Self-Organizing Map (SOM), is an efficient unsupervised classification method, that is very suitable for visualisation of multidimensional data. It involves two phases of processing: training and mapping, which is the main turning point comparing to clustering methods. In fact, in all the clustering methods, an input dataset is clustered into a certain number of the clusters. In SOM, however, a dataset is firstly employed for training in an unsupervised way, where each area is expected to respond to specific patterns. In the clustering methods the bordered are often hard, while in SOM there is no specific border, and the areas are approximately mapped. Another important differentiation point is that clustering methods are not often meant to involve a test phase.

Self-organizing map was firstly introduced by Teuvo Kohonen in the 1980s decade, as an efficient method for the clustering [83]. It is now mostly used for both classification and visualization. An appropriate 2D or 3D colour-coded graph, is commonly used for the visualization, which is called self-organizing map. The learning method is regarded as an artificial neural network, differing from Multi-Layer Perceptron (MLP) in the learning process. The learning process in self-organizing map is based on the competitive learning, that is against MLP, in which error is attempted to be minimized.

In this method, a certain number of the neurons, N, (N is a hyper parameter), are evenly arranged in a 2D or 3D setting according to the first 2 or three principal components, respectively. It is worth noting that the original learning method relied on assigning a set of the small learning weights to the neurons in a random way. Nevertheless, with the former alternative, the learning is by far faster, as the weights already receive a good initiation. For all of the training data, fed to the network, Euclidean distance of the sample and all the neurons weights are calculated. The neuron with the lowest distance of its learning weights to the sample, is considered as the most similar neuron to the sample, and named Best Matching Unit (BMU). The weights of the neurons, laying within a vicinity to the BMU, receive updates during the training, causing them to be pulled towards the training sample. The weight update for the neuron v is calculated as follows:

$$W_v(s+1) = W_v(s) + \theta(u, v, s)\alpha(s)(D(n) - W_v(s)) \qquad (3.31)$$

where s and t is the step index and the training data index, respectively. α is a monotonically decreasing function, employed to assure convergence of the learning weights. θ is the neighborhood function, between neuron v and neuron u in the vicinity of the BMU. Although in many applications a unity function can be used for θ, a Gaussian function is recommended in the majority of the applications. Depending on the implementation, s can imply on the recursion steps after including all the training samples indexed by n. It is important to note that the output of SOM is not a classification, as performed by the methods like MLP. Instead, SOM gives a mapping result, rather than the classes, which can be regarded as a golden feature for the visualization, or for the cases when the between-class borders are not sharp boarders. It is evidently preferred to use mapping rather than classification in such the cases.

3.3.3   Hierarchical Clustering

This method is indeed a derivation of K-Means clustering method. In this method, the learning process is started with a single centroid, calculated as the overall gravity centre. Then, a small random vector, "noise", is added to the centroid to create another centroid. The K-Means clustering is then, applied to the data with the new centroids, and the same procedure is repeated several times until a certain number of the clusters is reached. Although this is based on adding a noise to the overall centroid, therefore it so inherently encompasses a level of the uncertainty in the clustering, but is more stable than the K-Means, especially when the K-Means come with the random initialization.

Part II

Essentials of
Time Series Analysis

4

Basics of Time Series

The term time series, is defined by "a sequence of data in a known time order", and comprises a number of the notions and considerations, entailing further explanations. This chapter is dedicated to bring up the definitions and presumptions which need to be considered, when time series analysis is an objective. Sections of the chapter are arranged in an order from the basic introductory, up to the advanced definitions to addressing rather sophisticated topics such as sector processing, which have been recently introduced to this context. The first two sections, addressed in the introduction, try to bring the reader into the context for further elucidation as these notions will provide a base for the theoretical foundations of the upcoming chapters.

4.1 Introduction to Time Series Analysis

In the previous chapters, it was mentioned that image is not included as part of data that will be addressed throughout this book [105]. Regardless of the controversies which occur upon this subject, we assume that images cannot be considered as signals, and therefore an exclusion of image does not offend generality of the methods. A time series can be constituted of a set of the multi-dimensional feature vectors ordered according to the time occurrence of the vectors. Recall from Chapter 1, where signals and time series were linked to the phenomena creating them, such that variations of the phenomena can be explored by studying the resulting time series. It is in fact, the main goal of a great majority of the theoretical and practical attempts on time series analysis, is focused on extracting those informative contents of the time series, which are related to the phenomenon projected by the time series, either through identification, or by classification which is in turn considered an especial case of system identification [3][42][41]. An important topic of study, which extensively received interest from the scientists and engineers, is "system identification", which is entirely based on time series analysis. Time series analysis in this topic, and in most of the cases, refers to building a parametric model of the underlying system creating the time series, and the objective is concentrated on the methods for finding the model parameters [73][119][21][142]. The identified model of the system is employed for different applications. Estimating future contents of the time series is itself a prediction problem, or classification of the time series by using the model parameters, are two typical examples of such applications. Machine learning on the other hand, put the functional analysis of the systems into the learning objective, and hence deals with learning the system function instead of the parameters, by extracting information from the time series. This is indeed a case of system identification, implicitly performed in a non-parametric fashion. Hyperparameters of a machine learning method in the learning process, can be indirectly concurrent by the system parameters, even though these parameters profile the system function, where the system implies the underlying phenomenon creating the time series. In any case, the processing method attempts to learn dynamic contents of time series, or in another word, to extract parameters either from the system based on a reckoning model, or from the system function based on its behaviour at the output.

In a holistic view, there are three different ways to model dynamic contents of time series: deterministic, chaotic and stochastic model. These models will be addressed in the next section. Nevertheless, at this point, the essentials of how to formulate a model for time series analysis will be covered. For simplicity in model description, the time series is presumed to be single dimensional throughout this chapter. Generality of the upcoming formulation will not be lost by this presumption, as any multidimensional time series of dimension M can be decomposed to M single dimensional time series by considering each elements of the time series independently. Assuming that time series $X_i[t]$, $(t = -T, ..., 0, 1, ..., T)$, is a single dimensional sample from the training dataset of N samples $(i = 1, ..., N)$ (N is named the sample size), a system identification method corresponds to the methods for finding a mathematical model which is capable to describe current and the past values of the system, in addition to predict future values of the time series (value of the time series at $t = T + 1$). Contents of the time series in the present and past time are always invoked to predict future of the time series (time series values at $t = T + 1, ...$).

$$\hat{y}_i[t] = f(x_i[t-1], ..., x_i[t-T]) \tag{4.1}$$

It is obvious that dynamic contents of time series must be taken into account, to find the model parameters. It is often the case that the time series is sliced into the temporal windows of L samples, and the model parameters are found by utilising contents of each window, sliding over the time series with an overlap of V samples between each two successive windows. Root mean square of the predicted error, defined as the difference between actual values of time series and the predicted ones, can be employed as an informative metric to find an understanding about the model capability in identifying time series.

$$I = \sqrt{\frac{1}{L} \sum_{t=1}^{L} (y_i[t] - \hat{y}_i[t])^2} \tag{4.2}$$

$$i = 1, ..., N$$

where $\hat{y}_i[t]$ is the predicted value of the output $y_i[t]$ at the time, and $y_i[t]$ is the actual value of the output. In an ideal situation, where the system is perfectly identified by its parameters, the value of the time series on all points can be accurately predicted by using its present and pasts, and the I becomes 0.

In the derivation 4.1, there are three notations which require further attention: i for the sample number, t for the time variable, and T for the length of the temporal window which preserve dynamic of the system. The notation i addresses sample or sometimes the subject number, i.e., biomedical applications. This is in another language of the stochastic processing, termed by the ensemble index. As an example, lets assume we have N recordings of electrocardiograph from N patients, and here at this point, i refers to the subject number. It is obvious that t denotes time variable for a specific recording, e.g., sample i. T denotes the system order, or the extent to which the dynamic contents of the time series are supposed to be preserved. In parametric system identification, T is also regarded as the number of the model parameters.

Considering derivation 4.1 once again, three cases might happen:

• The number of the samples (N) is less than the number of the system order or model parameters T

• The number of the samples (N) equals to the number of the system order or model parameters T

• The number of the samples (N) is larger than the number of the system order or model parameters T

Imagine that the system parameters are found by solving a system of linear equations, in which the variables of the linear equations are the time series values at temporal points, the parameters are obtained by solving the equation system. The first case, is a situation, where the number of the equation is less than the number of the parameters. It is clear that there are an unlimited number of the solution for such the case, which should be obviously avoided. In the second situation, the number of the parameters is equal to the number of the equations. In this case, if the system equations are mutually independent, the system has a unique result. The last case, when the number of the equations is higher than the number of the system parameters, we face with an optimization problem, in which the prediction error, defined in Eq. (4.2), is tried to be minimised. One can easily see that in order to predict the time series, the value of the time series before $t = 0$ ($x_i[t] : t = -T+1, ..., 0$) must be known. These values are defined as the initial condition of the system. These initial conditions can effect on the output, and in a number of the methods, the rest of the conditions is defined as follows:

$$x_i[-T+1] = x_i[-T+2] = ... = x_i[0] = 0$$
$$i = 1, ..., N \tag{4.3}$$

It is important to know that number of the model parameters, T, as well as the overlap percentage, V, are regarded as two design parameters, optimised according to a prior knowledge of the time series nature, or by experimental procedures, or even by other theoretical considerations. They are sometimes found tentatively and empirically using a trial and error procedure.

4.2   Deterministic, Chaotic and Stochastic

Figure 4.1 depicts the three modelling categories.

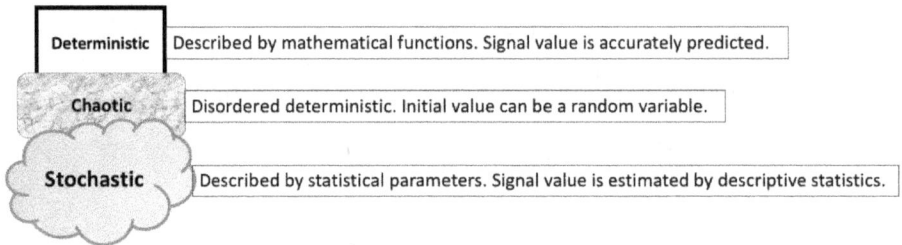

Deterministic	Described by mathematical functions. Signal value is accurately predicted.
Chaotic	Disordered deterministic. Initial value can be a random variable.
Stochastic	Described by statistical parameters. Signal value is estimated by descriptive statistics.

FIGURE 4.1: Three categories of modelling a system.

Recalling from Chapter1, where deterministic models were defined as the models in which time series can be expressed by a closed mathematical formula. In theory, a deterministic system is completely described by a linear time invariant differential equation. By definition, for a deterministic time series, a linear time invariant differential equation can be found, whose solution yields the deterministic time series. Practically phrasing, a deterministic system is interpreted as a system in which using the system equations of Eq. (4.1), with an identical number of the equations and parameters yields a time invariant set of the parameters for all values of t. This means that the system gives a unique mathematical formulation at its output, and thus the solution of the corresponding differential equation remains consistent over the time. It is not difficult to guess that such a system cannot be found in reality. In practical situations, even parameter of an ideal linear differential equation varies in time due to the environmental effects such as temperature and noise.

Chaotic systems on the other hand, are constituted of a nonlinear differential equation, with an unknown initial condition. In chaotic models, the output time series is the solution to a nonlinear differential equation with an unknown initial condition. Consequently, knowing the initial conditions, one can completely find the output time series, and that is why the chaotic models are sometimes called "disordered deterministic". This makes the initial conditions, critical for modelling the time series in the chaotic manner, and accounts for the turning point of such the models.

In stochastic models, an output time series is assumed to have resulted from a nonlinear and time dependent differential equations. We cannot fully define a linear time invariant differential equation for the system, so the solution of the system is not uniquely identified. The differential system equation describing the output, originating the time series, have a higher number of the equations than the system parameters. As a consequence, a complete solution cannot be found for the output, and therefore, an optimization technique must be invoked for the system identification. This book deals mainly with stochastic time series due to their practical and theoretical importance. In most of the practical situations, a stochastic model can be found to analyse the learning process, whereas the chaotic counterpart whose application is not seen as broadly as stochastic models.

4.3 Stochastic Behaviors of Time Series

A time series is often characterised by its values together with the variation of the time series value in time. Random variable theories are important elements for analysis of stochastic time series. In this section, parts of the random variable fundamentals are explained. Importance of such descriptions will appear in the future sections where different methods are described. It is worth noting that in statistical time series analysis, we encounter with two different types of variations: - Variation of the time series values for a certain subject with respect to time, t - Variation of the time series value at a certain time point with respect to the subjects, i.

In stochastic processing, average value of the time series i, is calculated by:

$$\overline{x}_i = \frac{1}{T} \sum_{t=0}^{T-1} x_i[t] \tag{4.4}$$

This is known as the first order statistic (moment) of the time series i. The reader must pay attention that the onset of the variable "time" (t) is the point $t = 0$ for most of the calculations of the temporal variables, in contrast with the subject number which starts from $i = 1$. The second order statistic (moment), is named "autocorrelation" function, and obtained by:

$$r x_i x_i(\tau) = \sum_{t=0}^{T-1} x[t]x[t + \tau] \tag{4.5}$$

Higher order statistics are calculated similar to the Eq. (4.6), which is a function of two variables.

$$C x_i(\tau_1, \tau_2) = \sum_{t=0}^{T-1} x[t]x[t + \tau_1]x[t + \tau_2] \tag{4.6}$$

Higher order statistics are known as "cumulants", e.g., third order cumulant. Third order cumulant has interesting characteristics. It is a linear operator and also its value tends to zero when the statistical distribution of the time series tends to the Gaussian shape. The higher order cumulants than 3 are not considered as the linear operators. The nomenclature of statistics in the stochastic processing is listed as follows:

- First order statistic: Mean value

- Second order statistic: Autocorrelation function

- Third order statistic: Third order cumulant

- Forth order statistic: Forth order cumulant

It is important to note that when considering the time parameter t in the statistics, the moments are functions of time τ or multiple parameters $(\tau_1, \tau_2, ...)$.

The addressed statistics are all calculated over time. In contrary, a number of the statistics are calculated over the population, or subjects, rather than the time span. This are named statistics or population-based moments. Average (μ) and variance (σ) value of a parameter are two well-known statistics defined over population, known as the first and the second statistical moments, calculated as follows:

$$\mu[t] = \frac{1}{N} \sum_{i=1}^{N} x_i[t]$$

$$\sigma_i^2[t] = \frac{1}{N} \sum_{i=1}^{N} (x_i[t] - \mu_i)^2$$

$$(4.7)$$

The reader should pay attention to this point that when the statistics are calculated over the population, the nomenclatures are preferred to be used differently to facilitate the understanding. A certain time series whose temporal variations remain stable for different time intervals is known as a stationary time series. This means that all the statistics of the time series (second, third,...) remain stable over different temporal windows. It is obvious that such a definition entails only theoretical value and cannot be found in the practical problems. A group of stationary time series whose temporal statistics are equal to their population statistics (counter moments) is known as ergodic time series. Presumptions like stationary and ergodic behaviours, drastically facilitate time series analysis, through the system identification, even though they cannot be seen in the practical problems. This will be partly addressed in the next section.

4.3.1 Cyclic Time Series

It is obvious that an stochastic time series cannot be periodic, easily since in periodic time series the future of the time series can be predicted. In reality, we face with a large number of the time series, whose values resemble repetitive contents, but cannot be categorized as periodic time series. In these cases, even though regularity cannot be observed in the value of time series, meaning that the values are not exactly repeated at a certain, priory known points of the time. Nevertheless, certain patterns are repetitively seen over a time span. These types of time series are named as cyclic time series. A recording of electrical activity of heart, so called electrocardiogram, is considered as a cyclic time series [33][70][132]. One should consider that the cardiac cycle of a person is not a fixed number and varies, because of the physiological conditions such as respiration phase (inspiration or expiration), biochemical characteristics of the blood, hormonic activities of the body, stress (mental or physical), and many other factors which are not, yet, understood [121]. This electrical activity initiates a mechanical activity in the heart that yields blood circulation in the body. Recording of mechanical activity of heart, produces a cyclic timer series of the acoustical signal, named phonocardiograph. This is another cyclic time series which resembles even higher level of randomness comparing to the electrocardiogram. Figure 4.2 sketches a few cycles of a phonocardiograph signal:

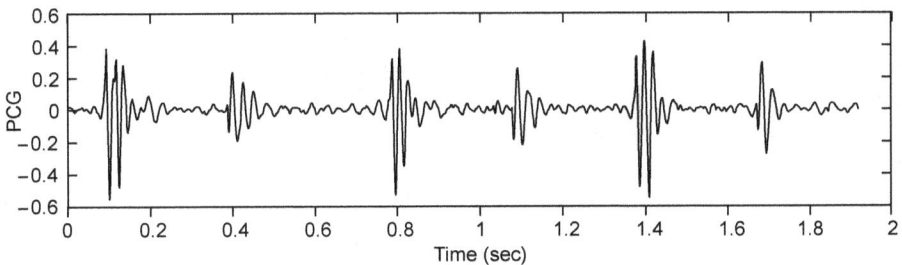

FIGURE 4.2: Three cycles of heart sound signal, known as phonocardiograph. Cyclic behavior with variant cycle is clearly seen in the graph.

Recording of lung sounds, or other natural phenomenon like the temperature at a certain geographical points of the earth all show a cyclic variation. Here, random variable theory can profoundly help for classification and prediction problems.

A cyclic model for processing a time series assumes that the time series resembles random behaviour within the cycles, and also the cycle duration by itself is also a random variable. A presumption of the cyclic time series, is that the onset and the endpoint of the cycles are priory identified by a separate source of information out of the time series contents. In many practical cases, an auxiliary signal is recorded along with the time series synchronously. The auxiliary signal helps to identify the onset and the endpoint information of the time series. Such a situation is widely seen in industrial and medical applications. In the following sub-sections, some of the definitions, that will be invoked for the learning process will be described.

4.3.1.1   Sector Definition

By definition, a sector of a cyclic time series, is a set of the temporal frames of a cyclic time series, all starting at a certain relative distance and having a certain relative length, with respect to the underlying cycle. It is obvious that cyclic duration is not a fixed number, and therefore a temporal frame with a fixed relative length of certain ratio with respect to the cycle, cannot have a fixed temporal length. This is the case for the initial point of the frames. If the initial point is a calculated based on a certain ratio of the cycle, its position within a cycle depends on the cycle duration and is not temporally fixed. As we will see in the later chapters, sector definition can efficiently facilitate learning process. It is important to bear in mind that a sector length as well as the offset point (from the beginning of a cycle) are defined based on a fraction, or even a fraction, of the cycle length [53].

4.3.1.2   Uniform Sectors

A set of the sectors with identical length within a cycle, is defined as the uniform sectors. One should note that the length of the sectors within one specific cycle is meant by this definition. It is obvious that two different cycles encompass sectors with different lengths due to the different cycle durations. The uniform sectors, divide each cycle into K temporal frame, each having the same length. The K is fixed for all the cycles and regarded as a design parameter. Considering the time series $x_i(t)$ with L_i number of the cycles, a set of the K uniform sectors is defined as follows:

$$x_l(t, k) = x_l(\tau_k + t)$$
$$k = 1, ..., K$$
$$l = 1, ..., L_i$$
$$\tau_k = (k - 1)\frac{T_l}{K} \tag{4.8}$$
$$t = 0, ..., \frac{T_l}{K} - 1$$

where k is the sector number ($k = 1, ..., K$), l is the cycle number ($l = 1, ..., L_i$), T_l is the duration of the cycle l, and τ_k is the offset of the sector with respect to the cycle duration.

4.3.1.3   Growing-Time Sectors

The sectors within a cycle can have a different length, and especially one interesting option is to choose the sectors with the growing length. For a set of the sequential sectors within a cycle of a cyclic time series, the sectors are considered as the growing-time sector if the two following criteria are fulfilled:

- For each two successive sectors, the length of the first sector is shorter than the length of the second sector

- For each two successive sectors, the second sector completely includes the first sector

It is evident that contents of the first sector appears in all the consecutive sectors, and as we will see later in the sequels, this sector receives the highest importance by the learning methods [53]. By this definition, one can bring a number of different possibilities of the growing-time sectors through a heuristic fashion, however, this book describes only three well-defined growing schemes which will be employed later in different learning processes:

- Forward growing-time sector

- Backward growing-time sector

- Bilateral growing-time sector

A forward growing-time scheme for a set of the sectors is the scheme in which the onset of the first sector is exactly coincident with the onset of the cycle, and is common for rest of the sectors within a cycle. In this scheme the first sector begins at the beginning of the cycle, and its length grows in time, within a cycle, until it covers the whole cycle:

$$x_l(t, k) = x_l(t): \quad t = 0, ..., k\frac{T_l}{K} - 1$$
$$k = 1, ..., K \tag{4.9}$$

where $x_l(t)$ is the contents of the time series for cycle l, $x_l(t, k)$ is contents of the time series for the sector k and cycle l, and $T - l$ is the length of the cycle l. One should be aware that for each sector k, the onset point is $t = 0$. The total number of the sectors per cycle K is a design parameter which makes implications on the growing rate of the sectors as well.

A backward growing-time scheme for a set of the sectors is the scheme in which the **endpoint** of the first sector is exactly coincident with the endpoint of the cycle and is common for all the sectors of a cycle. In this scheme, the first sector lays at the end part of the cycle with the endpoint coincident at the endpoint of the cycle, and its length grows reversely towards the beginning of the cycle until to cover the whole cycle:

$$x_l(t, k) = x_l\left(t + T_l - k\frac{T_l}{K}\right), \quad t = 0, ..., k\frac{T_l}{K}$$
$$k = 1, ..., K \tag{4.10}$$

In bilateral growing-time scheme, the first sector lays somewhere within the cycle, and grows from the both ends until to cover the whole cycle. The center of the first sector is called "**GROWING CENTER**", T_G, defined by a percentage of the cycle duration:

$$T_G(l) = \eta \frac{T_l}{100} \tag{4.11}$$

where l is the cycle number. η is a percentage of the cycle at which the center of the first sector sets. Contents of the sectors for a bilateral cyclic time series can be formulated by:

$$x_l(t, k) = x_l \left(t + (1 - \frac{k}{K})T_G(l), \quad t = 0, ..., k\frac{T_l}{K} - 1 \right) \tag{4.12}$$

$$k = 1, ..., K$$

Obviously, the growing center is not obligated to be at the middle of the cycle, but the growing rate is identical for the both ends of the sector. This doesn't imply that the temporal length **equally** grows from the both ends.

It is easy to see that regardless of the scheme of the growing-time, (1) the first sector is the shortest sector, (2) the last sector covers the whole cycle, and (3) the total number of the sectors for a cyclic time series i is KL_i in which L_i is the total number of the cycles existing in the time series. These are three important characteristics of the growing-time sectors.

4.3.2   Partially Cyclic Time Series

A cyclic time series whose statistical characteristics remain almost consistent within certain segments of a cycle, is known as segmented cyclic time series [55]. Figure 4.3 shows two cycles of a phonocardiograph (top) along with the electrocardiograph (bottom). It is clearly seen that in some, but not all, segments of a cycle, the phonocardiograph shows stochastic behaviors.

In a certain segments of a cycle, for example the segments denoted by S1 (the first heart sound), the time series behaves differently comparing to the segment denoted by the diastolic murmur or the systolic murmur. Clearly, learning the first heart sound segment demands a different learning method with respect to the diastolic part. Such the time series are therefore, categorised in a different group, to allow flexibility in introducing appropriate models for the learning process which will be discussed later in in the upcoming chapter.

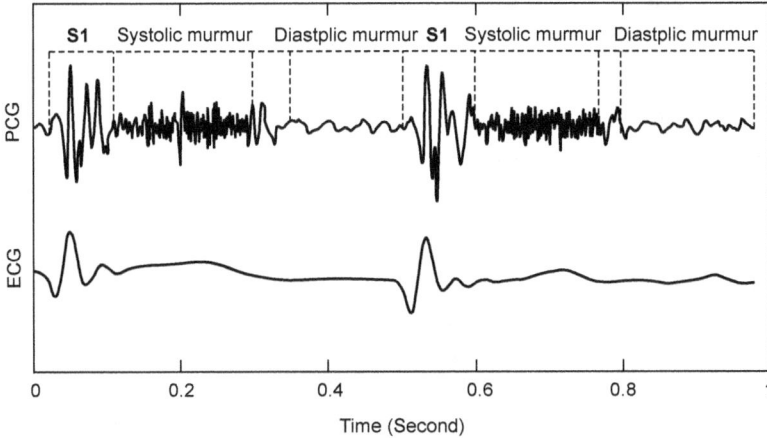

FIGURE 4.3: Two cycles of phonocardiograph (top) and electrocardiograph (bottom). Segmental cyclic behavior with variant cycle is clearly seen in the graph.

4.4 Time Series Prediction

In practice, most of the time series represent stochastic characteristics, and clearly the future value of the time series, cannot be precisely predicted. As a result, there is always an error associated with the prediction of a time series value. It is a main assumption in many attempts, and in many aspects of time series analysis including prediction, that the phenomenon behind a time series, is identifiable, and the random behaviors of the results obtained from the analysis are caused by the noise or any other reckoning element. In many problems of system identifications, it is deemed for the prediction error to resemble a Gaussian statistical distribution, whose variance and mean values denote power and bias of the noise, respectively. In order to predict time series value at a certain point of time, there are fundamentally two different strategies: machine learning-based and model-based strategies. The first strategy, involves a machine learning method to learn variations of the time series, and then prediction of the future values of the time series is performed by using the machine learning model which was previously trained using a specific training method. Various machine learning models and training methods are available nowadays. Details of this strategy is discussed in a future chapter of the book (Chapter11, Recurrent Neural Network), where dynamic neural networks will be presented. The second strategy, is based on a process of system identification using a specific model, by which the future values are predicted. The model can be identified by any kind of the linear or nonlinear manners, selected according to the time series behavior and estimated based on the prediction error. Nonlinear and linear models, each entails advantages and disadvantages against the other. One of the most widely-used methods, is based on a kind of linear model, named "AutoRegressive Moving Average" (ARMA), model in which it is assumed an input time series is resulted from a linear system with the following differential equation:

$$b_0 y_i[t] + ... + b_{T-1} y_i[t - T + 1] = a_0 x_i[t] + ... + a_M x_i[t - M + 1] + \varepsilon_i \quad (4.13)$$

Then, each point of the time series is resulted from the system output added by a white noise ε_i acting as a random variable. Here, we are faced with a problem of system identification, in which the parameters a_l, b_k : $l = 1, ..., M$, $k = 1, ..., T$ are to be found by the identification process. In the above parametric system modelling, if $\forall l > 0 : a_l = 0$, the system is named "AutoRregressive" (AR) model, in which only time shifted outputs as well as the input at $t = 0$ appear in the model for calculating the output, without having the shifted inputs. If $\forall l > 0 : b_l = 0$, then the system is named "Moving Average" (MA) in which only the shifted inputs contribute to the output without having the shifted outputs.

In order to simplify the process, an AR model is firstly employed to find the parameters $\langle a_i \rangle$. The parametric model assumes that the estimation error is calculated using real outputs $y_i[t]$ and the estimated ones. From Eq. (4.13) with the above assumptions we have:

$$\hat{y}_i[t] = \frac{1}{b_0} (a_0 - y_{i-1}[t] - ... - b_{T-1} y_i t_T + 1 - \varepsilon_i) \quad (4.14)$$

and the minimum least square error is obtained:

$$E = \left(y_i[t] - \hat{y}_i[t] \right)^2$$

$$= \left(y_i[t] - \frac{1}{b_0} (a_0 - y_{i-1}[t] - ... - b_{T-1} y_i t_T + 1 + \varepsilon_i) \right)^2 \quad (4.15)$$

$$\frac{\partial E}{\partial b_l} = 0$$

Taking the autocorrelation function from the both sides removes the noise term which yields the parameters by:

$$B = [b_0, ..., b_{T-1}] = R^{-1} \underline{r} \quad (4.16)$$

where

$$R = \begin{bmatrix} r_0 & r_1 & r_2 & ... & r_{T-1} \\ r_1 & r_0 & r_1 & ... & r_{T-2} \\ r_2 & r_1 & r_0 & ... & r_{T-3} \\ ... & ... & ... & ... & ... \\ r_{T-1} & r_{T-2} & r_{T-3} & ... & r_0 \end{bmatrix} \quad (4.17)$$

$$r_l = \frac{1}{T} \sum_{k=0}^{T-l-1} y_k y_{k+l}$$

The above correlation matrix is positive and symmetric and definite, so the inverse matrix exists. However, in the digital implementations, there are various methods for the calculation in recursive form which is always favored to the digital processors. Interested readers can find such implementations in [22][104]. Implementation of the moving average model is similar to the AR model, and so is the ARMA model. Details of such implementations can be found in [22]. Description of nonlinear modelling for system identification is well-beyond the scope of this book, and the interested readers can refer to the pertinent literature in system identification.

4.5 Time Series Classification

In time series analysis, classification is defined as establishing an association to a time series, based on the prior knowledge about the possible cases of the time series. From the previous sections, it is not difficult to realise that time series attributions like stationary and ergodic refer to absolute definitions and cannot be found in reality. Nevertheless, such the presumptions can help to establish statistical methods for exploring time series characteristics, and by this means to extract informative contents from the underlying phenomenon. To this end, contents of time series can be employed to associate the time series with a specific class of that phenomenon. An example of such a process, is a time series of physiological activity inside the body such as brain activity. The resulting time series can be recorded from several points over the scalp, and processing of this set of the time series, named electroencephalogram (EEG), can reveal important information about the brain activity which is considered as the phenomenon in this example. Classifying the time series of EEG can help to find out type of the brain activities, such as sleep phase, or even to detect a number of the brain defects like epilepsy. Consequently, time series classification is indeed interpreted as a way of obtaining an understanding about the underlying phenomenon. This is certainly a case of system identification, however, formulated in another manner when it comes with machine learning. In this context, the phenomenon behind the time series is not deemed to be identified by itself, but instead its manifestations are learned by the corresponding methods. Classification was expressed as establishing an association to a time series, based on the prior knowledge about the possible cases of the time series. It is obvious that time series classification is tightly linked to learning dynamic contents of the time series. In this sense, there are basically three different models for time series classification, depicted in Figure 4.4, and categorised according to the underlying fashion for preserving the dynamic contents in the learning process:

Learning Time Series Dynamics		
Feature Level	**Input Nodes**	**Classifier Architecture**
• *Multi-Layer neural Network* • *Vector Quantification* • *...*	• Time Delayed Neural Network • Time Growing Neural Networks • ...	• Recurrent neural Networks • Hidden Markov Model • ...

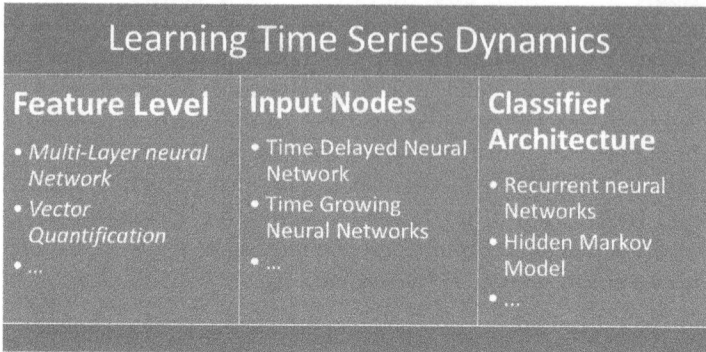

FIGURE 4.4: Dynamic contents of time series can be learned in three manners: at the feature level, at the input node, and in the entire architecture of a classifier.

- Preserving dynamic contents of time series at the feature level in conjunction with a static classification method

This model has been introduced as a basic classification method in artificial intelligence. Temporal variations of time series are considered at the feature level, such as using the spectral energies or time frequency decomposition, along with a multi-layer perceptron neural network. In such the models, art of the classification resides in extracting powerful features, and the classifier can use any kind of the static method such as support vector machine [122]. This model has been studied in several investigations on natural speech recognition, biological signal processing, and ultrasound detection. Different methods of frequency analysis, time analysis, linear time-frequency decomposition, and nonlinear time-frequency decomposition have been employed as the mathematical tools for the feature extraction [135][125][150].

-Preserving dynamic contents of time series in the architecture of the classifier

In this model, the main focus of the classification lies in the architecture of the classifier, that receives the time series values either single by single, or as a pack of the multidimensional feature vectors, and performs the classification according to the received contents. Recurrent neural networks can be referred to as one of the most used models for this purpose, which will be separately described in later chapters. It is worth noting that the dynamic contents of the time series is learned by a fixed architecture, but variant values, for the parameters in time according to the time series values [108][112].

- Preserving dynamic contents of time series in both the classifier architecture and at the feature level

In these models, parts of the learning process is performed at the feature level, which can in turn contribute in the input part of the classifier architecture, and the rest of the learning is followed by a dynamic architecture.

Here, it is reemphasized that dynamic architecture implies a fixed structure, but with time varying values of the parameters, whose values depends on the past and present contents of the time series [108][112]. Hidden Markov model, hybrid methods and recurrent time-delayed neural networks are categorised at this group, which will be explained in the upcoming chapters.

5

Multi-Layer Perceptron (MLP) Neural Networks for Time Series Classification

In the functional point of view, a multi-layer perceptron neural network inherently performs a nonlinear regression with the difference from classical methods where the parameters associated with the method are found in an iterative way instead of analytic ways. Any iterative method needs terminating criteria, to terminate the recurssion, when the criteria are met, otherwise the method will make an over-fitting on the decision boundaries. The number of the epochs for training in conjunction with the classification error, are the two criteria employed in most of the applications. The architecture of a MLP contains an input node, constituted of the feature vector, followed by a number of the layers, named hidden layer, along with an output layer (see Figure 3.8).

Each of the hidden and output layers contains a number of the neurons. There is no analytic way to calculate an optimal value for the number of the neurons, required to perform the regression, at different layers, likewise for the number of the hidden layer. However, in most of the practical situations, one hidden layer is sufficient, and the number of the neurons in that layer depends on the training variety. It has been experimentally shown that increasing the number of the neurons after a certain range at the hidden layer of a three layers MLP, will put negligible effect on the classification performance, which in turn depends on the training data size [48].

It was previously discussed that MLP itself cannot preserve dynamic contents of time series in its architecture, and therefore for the problems of the classification kind, there is a need for further elaboration. Even though, these contents can be considered in the feature space by employing different sorts of time-frequency representations such as wavelet transformation techniques, incorporating the temporal variation of an input time series into the architecture of a MLP can elaborate the classification quality [120][122][155]. It can also assign further flexibility to make the trade-off between the feature level and the classification level for analysis in which subtle variations are required to be considered. This issue has been explored by different scenarios, e.g., time-delayed neural network [143], time growing neural network and recurrent neural network. The first two alternatives will be explained in this chapter and the later is postponed for Chapter 11.

5.1   Time-Delayed Neural Network (TDNN)

Time-Delayed neural network comes with an offer for architecture, in which the input node has the capability to include dynamic contents of time series [106][128][134]. By means of this architecture, subtle variations are leaned by the input architecture of the classifier, and the short-time variations can be preserved by the feature vectors. Figure 5.1 illustrates architecture of a typical TDNN:

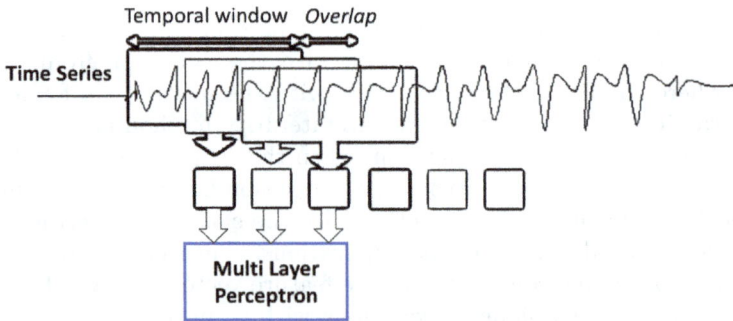

FIGURE 5.1: General architecture of time-delayed neural network.

One way to learn short-time information of time series, is by using frequency contents of the time series, $x_i[t]$ within the interval of length T starting at $t = 0$. Spectral energy of the signal over the frequency band, (ω_1, ω_2) is estimated based on Fourier analysis:

$$P_i(\omega_1, \omega_2) = \frac{1}{T} \sum_{\omega=\omega_1}^{\omega_2} \left| \sum_{t=0}^{T-1} x_i[t] W[t] e^{-j\omega t} \right|^2 \tag{5.1}$$

In this formula, the shape of the window is characterised by $W[t]$ to mitigate effect of the side lob and frequency leakage [132][32]. Hamming, Hanning, Gaussian and Kaiser windows are the most common shapes for the spectral calculation. Sometimes an overlapping between successive windows are employed in order to better compromise between time and frequency resolution. Interested readers can refer to [104] and [22] for obtaining profound intuition about the signal processing details and the theoretical description of this topic. Nevertheless, the mathematical formulation is given for the case without temporal overlapping. The interested readers can easily expand it for the overlapping windows. In Eq. (5.1), the contents of the time series can be calculate for the temporal window Δ, as follows:

$$P_i(\lambda, \omega_1, \omega_2) = \frac{1}{T} \sum_{\omega=\omega_1}^{\omega_2} \left| \sum_{t=0}^{T-1} x_i[t + (\lambda - 1)T]W[t]e^{-j\omega t} \right|^2 \tag{5.2}$$

$$\lambda = 1, ..., \Delta$$

where Δ is the number of the temporal windows. For simplicity in formulation, a frequency band is defined as follows:

$$B \doteq \{b = (\omega_1, \omega_2) : \omega_1, \omega_2 \in Z^+ \wedge \omega_1 < \omega_2 < \omega_{max}\} \tag{5.3}$$

The spectral energies, derived by Eq. (5.2), are calculated over M frequency bands, which yield:

$$P_i(\lambda, b_l) = \frac{1}{T} \sum_{\omega \in B} \left| \sum_{t=0}^{T-1} x_i[t + (\lambda - 1)T]W[t]e^{-j\omega t} \right|^2$$

$$\lambda = 1, ..., \Delta \tag{5.4}$$

$$l = 1, ..., M$$

$$\forall l \ b_l \in B$$

The input node of a TDNN is constituted of the spectral energies calculated over M frequency bands, for Δ temporal windows. The total number of the features at the input node is ΔM features of spectral energy, yielding the following feature vector:

$$\mathcal{P}_i = \left[P_i(1,1), P_i(1,2), ..., P_i(1,M), P_i(2,1), P_i(2,2), ..., P_i(M,\Delta) \right] \tag{5.5}$$

Given the weights of the neural network at the hidden and the output layer by $\{W_{j,k}\}$, each neuron of the output layer of the TDNN gives:

$$\phi_j(X_i) = g_2(W_{2,j}^T \cdot g_1(W_{1,k}^T \mathcal{P}_i - \theta_k) - \theta_j) \tag{5.6}$$

The weights together with the thresholds $\langle \theta \rangle$ are found by back propagation error method, as described in Chapter 3. The window length as well as the possible overlapping percentage are considered as the hyperparameters, which can be found either empirically, or tentatively, or systematically through an optimization process, however, sometimes prior knowledge about the time series nature is invoked to find a set of the optimal parameters. Another set of parameters is the number of the frequency bands M, which are obtained intuitively at this point. As you will see in the future chapter, certain sophisticated deep learning methods can help to find these bands [122]. The detailed description of this method is found in Chapter 7, when deep learning methods and the cyclic time series are fully described.

5.2   Time-Growing Neural Network (TGNN)

From the previous section, it can be easily understood that TDNN is indeed a type of MLP with the difference that in TDNN, temporal variation of a time series is included at the input node of the neural network, keeping the rest of the architecture the same as an MLP. The training method is also identical to a simple MLP. Likewise to the TDNN, there is another neural network in which dynamic contents of an input time series are preserved at the input layer and the rest of the process is similar to MLP, in terms of both the architecture and the training method. This new neural network is named Time Growing Neural Network (TGNN) [48][54]. Input node of a TGNN is constituted of spectral contents of the input time series, calculated over several overlapping temporal windows. The length of the windows grow in time, with a growing rate of K, until covering all of the time series length. The first node corresponds to the shortest window, the second temporal window covers the first one in addition to some more samples of the time series, and so on until the last window which includes the time series entirely. Figure 5.2 depicts the general architecture of a TGNN:

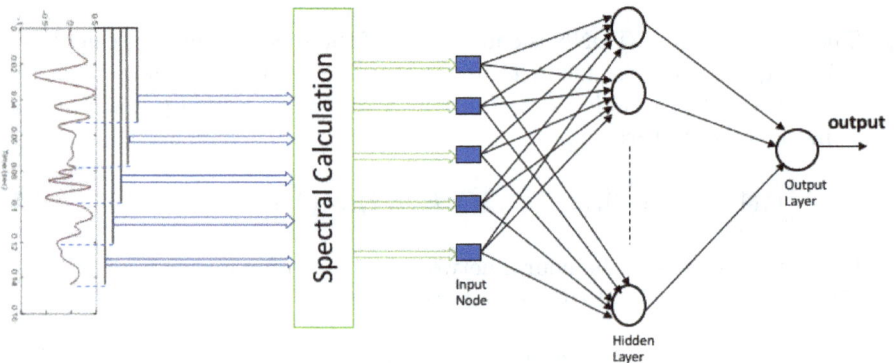

FIGURE 5.2: General architecture of a forward time growing neural network.

Each window can be characterised by its spectral contents, calculated using periodogram. Lets assume that the length of the first window's length and the growing rate is T_0 and G, respectively. Spectral contents of the temporal windows l can be calculated using periodogram:

$$X_i(\omega_1, k) = \frac{1}{T_0 + kG} \left| \sum_{t=0}^{T_0+kG-1} x_i[t]W[t]e^{-j\omega t} \right|^2 \qquad (5.7)$$

Assuming that a set of the frequency bands, defined in Eq. 5.3, is employed to calculate the spectral energies, the input node is similar to the one with TDNN, but with the difference in calculating the spectral energies:

$$P_i = \left[P_i(1,1), P_i(1,2), ..., P_i(1,M), P_i(2,1), P_i(2,2), ..., P_i(M,\Delta) \right]$$

$$P_i(k, b_l) = \frac{1}{T_0 + kG} \sum_{w \in B} \left| \sum_{t=0}^{T_0 + kG - 1} x_i[t + (k-1)G] W[t] e^{-j\omega t} \right|^2 \tag{5.8}$$

$$\forall l : b_l \in B$$
$$l = 1, ..., M$$
$$k = 1, ..., \Delta$$

where B is defined in Eq. (5.3).The hidden layer and output layer, both together, map the input node to another domain by the nonlinear functions g_1 and g_2 as:

$$\phi_j(X_i) = g_2(W_{2,j}^T \cdot g_1(W_{1,k}^T P_i - \theta_k) - \theta_j) \tag{5.9}$$

The weights together with the thresholds $\langle \theta \rangle$ are found by back propagation error method, as described in Chapter 3.

The length of the first window T_0, namely the initial frame length, along with the growing rate G, are considered as the hyperparameters found through an optimization process. The window's length can grow in different ways, named the growing scheme, and will be formulated in the next section. Finding an appropriate set of the frequency bands, is a case of the learning at the deep layer of the network architecture which will be described in Chapter 7.

5.3 Forward, Backward and Bilateral Time-Growing Window

Forward TGNN (FTGNN), is a scheme of the growing window in which the windows all have an identical beginning point, but different ending points, each is longer by G samples than the previous window. A windowed signal is hence, defined by the following derivations:

$$\{x_i[t] : t = 1, ..., T_0 + (k-1)G , \ k = 1, ..., \Delta\} \tag{5.10}$$

In this scheme, contents of the first window, which appear at the beginning part of the time series, exist in all the other windows, and therefore, the neural network learns variations of the time series with respect to this window. Forward time growing neural network is especially efficient when we deal with the problem of learning dynamic contents of a time series by taking an initial segment of the time series as the reference. As such the behaviour occurs in the human auditory system. Let's imagine a short length acoustic signal of a type speech with a length of 1 second, stacks in a human ear of a healthy person. The auditory part of the brain receives the information and interprets the information, sequentially taking the first part of the information as the reference [32]. In fact, the first segment of the audible sound (typically with a length of 20 millisecond), remains as a residue until the following segments. Forward TGNN can serve as a model for the human auditory system. The corresponding parameters are the initial window length (T_0) and the growing rate (G), which should be selected appropriately.

Backward TGNN is a scheme in which the input node is composed of the contents of a set of the temporal windows, where the first window locates at the end of the time series. The rest of the windows grow backwardly with the growing rate G, until reaching the beginning of the time series. The temporal windows all have an identical ending point, but different beginning points. The beginning point starts from the last part of the time series with the length T_0 for the first window, and the zero point of $t = 0$ for the last window.

$$\{x_i[t] : t = (\Delta - k)G, ..., T_0 + \Delta G , \ k = 0, ..., \Delta\} \tag{5.11}$$

In the bilateral growing TGNN, the shortest window resides somewhere within the time series, and grows bidirectionally from the both ends, its beginning and its ending points, until covering whole the time series. The centre point of the shortest segment, is named the growing centre.

In the practical perspectives, all the three schemes of TGNN are common at this point that the shortest window is assumed to be a part of the time series which is roughly consistent over the subjects of the training data, and information of the rest of the time series is by far another variant, conveying the between-class information. This will direct us to the point to find the growing centre. Assuming that spectral energy of the temporal windows is calculated over the entire frequency band of the time series for all the time-growing scheme. For simplicity, we consider a forward time-growing scheme as in Eq. (8.2). Without losing generality of the theory, similar derivations can be driven for other schemes of the time-growing windows. The segment which provides the most stability could be used as the first window:

$$K_{Opt} = \arg \max_k \{P_i(k, b_l)\} \tag{5.12}$$

The K_{MAX} is indeed the index of the temporal window with optimal segregation.

This was one alternative to define the growing scheme. Nevertheless, any reader can contribute to define other heuristic alternatives depending on the case study. Use of wavelet transformation as an efficient tool for finding the growing centre has been reported in the recent studies, which will be discussed in Chapter 8.

5.4 Privileges of Time-Growing Neural Network

Use of time growing windows puts a positive impact on the classification performance, when it comes to preserving the dynamic contents of time series, and by this means serving as an elaborative classification method. These impacts have been experimentally investigated and compared to the two other alternatives, Time-Delayed Neural Network (TDNN) and Multi-Layer Perceptron neural network (MLP), in a number of the studies on classifying time series of physiological signals [48][53]. In this section these important characteristics of TGNN are investigated in an intuitive way, as well as compared to the other two alternatives, TDNN and MLP. The readers must bear in mind that all the properties described below remain valid when the training phase is ideal, without having been stuck in training bottlenecks such local minimums, or over-fitting. Cross-validation is a well-known technique to overcome such problems. In this technique, the training is stopped and the resulting performance is saved, and the same training is repeated several times, yielding the ultimate accuracy as the average of the accuracies resulted from the estimations during the cross-validation process. An important assumption to all of the following descriptions, is to perform an appropriate cross-validation with sufficient number of the iterations, making the conclusive terms fully subjected to an appropriate cross-validation. This point is not repeated in the rest of the descriptions, taken as a necessary presumption to all of the following conclusions. Another important assumption, which should be taken for the rest of this chapter is focused on the feature vectors, which are constituted of spectral energies. For other types of features, part or all the following points might be inconclusive.

5.4.1 TGNN includes MLP in its architecture

Referring to Figure 5.2, one can easily see that the largest temporal window covers whole the time series, and consequently, becomes a MLP where the temporal contents are included in the feature domain only. The rest of the windows help the neural network to learn dynamic of the time series.

5.4.2 TGNN can include TDNN in its structure

Assuming that the feature space is constituted of the spectral contents of training time series. It is known that there are different methods for estimating frequency contents of a stochastic time series [22][104], each having pros and cons, by the way for simplicity, the direct method of Fourier transformation is chosen to show pervasiveness of TGNN compared to the other alternative, TDNN. Assuming a forward scheme of the growing windows are used for time series characterisation, the second temporal window covers the first window in addition to the frame growth with the rate G. In the spectral domain contents of this window equal to the contents of the first window added by the contents of the added part as well as a cross modulation term which is negligible for the stochastic signals. Therefore, the value of the added window participates in the output, and consequently in the learning weights are assigned appropriately during the training phase.

5.4.3 TGNN is optimal in learning the first window

Along with the previous comparisons, here, the term optimal is employed to provide a comparative conclusion with respect to the other presented alternatives: MLP and TDNN. The readers should not forget that performing cross-validation, is obviously regarded as "a must" to all the conclusive terms of this section. Another important point to be considered is that all the comparisons, made in this section, are based on using identical feature vectors for all the three neural networks, and merely puts input nodes of the three neural networks into a comparative challenges. It is clear that MLP is incapable to include dynamic of time series in its architecture, and the learning weights are assigned to the whole time series. As a result MLP is not considered in comparison to TGNN in which the dynamic contents of time series are preserved at the input node. Contents of the first window, the shortest window, appear in all the other windows. Since a learning weight is automatically assigned to each temporal window, and the contents of the first window is included in all the other windows, the first window receives the highest number of the learning weights, and after that the second window, and so on, such that the longest window receives the least learning weight. One can easily see that a TGNN assigns the highest number of the learning weights to the first window (the shortest window). A consequence of such an inconsistent weight assignment, with the highest bias to the first segment, is better learning for that segment, since a miss-training of this segment effects the learning weights of all the other segments. This implies that the error of the neural network is far more magnified for the first segment as compared to the other segments, and consequently is biased towards providing a better learning for the first segment. For the TDNN case, the learning weights are consistently assigned to all the segments and therefore the segments are learned equivalently.

In many practical situations, it is desirable to explore if contents of a certain part of time series abruptly changes with the certain trend of variation. As TDNN cannot pay especial attention to the first segment, and the segment contents are non-stationary, the learning process will not converge to a reliable result. In contrary, TGNN can offer a much better classification performance. Experimental studies showed that classification of systolic click sound, the sounds heard because of the abrupt changes in the frequency contents, can be rather elaborated by using TGNN compared to MLP and TDNN [48][62].

6

Dynamic Models for Sequential Data Analysis

A time series, as a set of the sequential data points with a certain time order, typically carries information in its dynamic alterations over the samples. It is often seen that the samples are sequentially ordered in time, as expected from the term time series, which makes the notion dynamic contents meaningful in this context. The difference between sequential data and time series, lies at this point that in the time series case, the data sequence is ordered with respect to time, elapsing at a certain priori-known rate, whereas in the sequential data case, the temporal rate might not be consistent. This is in fact, the way of looking at data. Another important point to be noted, is the fact that time series is often either recorded, or obtained, however, sequential data can be attributed by a manipulator, method, or even a system, through a process named either "labelling" or registration. To put this point into a better perspective, let's consider a time series of an electrocardiogram signal, expressed by its peaks and waves, as depicted in Figure 6.1.

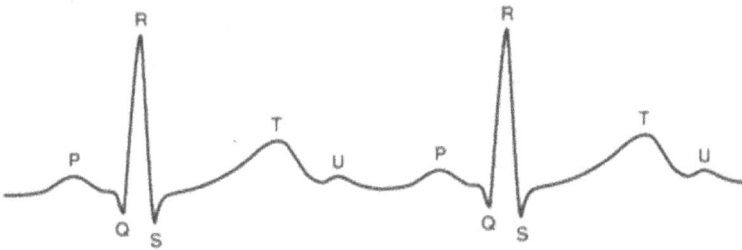

FIGURE 6.1: Tow cycles of an ECG signal, expressed as a time series with the known sequences of P-wave, QRS complex, and T-wave each conveying important information of electrical activity of a heart.

The data point shows values in time, measured from the body, constitute a time series, while assigning specific names on the peaks and waves, i.e., P-Q-R-S-T, leads to a sequence of the data. These peaks and waves, carry indicative information about electrical activity of heart, which is initiated from the pace-maker cells at the atrium and propagates towards the ventricles, and further down to the apical point of heart, and returns back to its initial point to create a self stimulating activity. This function can create a rhythmic sequence of contraction and relaxation, and both together establish blood circulation. A disorder in the electrical activity of heart creates abnormal alteration of the mentioned peaks or waves, such that studying these dynamic characteristics provides a way to explore electrical function of heart [132][70]. The peaks and waves are sometimes, and even in some cardiac cycles, disappear due to physiological or pathological conditions, and therefore might exhibit irregularities in their appearance frequencies. The ECG samples are however, present regardless of the cardiac condition. The difference between sequential data and time series has been addressed by this example, because of its role in extracting dynamic sequence from the data. Both the time series analysis and the sequential data analysis are mostly discussed for the purpose of extracting information. This is performed by presuming a certain model, postulated over a data group, in many scientific approaches. Here, it is worth noting that models are almost always employed when there exists a scientific or a technical hypothesis behind. A model is based on a set of the hypotheses, and differs from structures or graphs, in which hypothetical bases are not necessitated. The presumptions and hypotheses can be completely different for the time series analysis and the sequential data analysis. A very rough view suggests that a higher level of stochastic characteristics is associated with time series analysis rather than sequential data analysis. This can be clearly seen in the related graphs in which variations of the signal value resembles a stochastic time series, while the peak sequences are further deterministic!

It is understandable as the data in this case is assigned to the peaks and waves, in contrast to the time series value which is measured, always under the circumstance of a noise.

A delicate point always arises with the term "stochastic model". In fact, the world "model" usually carries a deterministic notion behind, coming from the absolute understanding about one or a set of the hypotheses. This is in contradiction with the term "stochastic". Perhaps, a better term would be "model for stochastic data". Nevertheless, we use the term "stochastic model" instead, for the simplicity. The readers must be aware of this naming and intention of the authors. This chapter introduces three well-known models for time series analysis. The presenting models are capable of preserving dynamic contents of time series. Although the name of the chapter can be somehow misleading, both time series and data sequences can be modelled by the models presented below. The first model, which will be described in the next section, is dynamic time warping. This model considers a certain sequences of the data, trained in the learning phase. The model involves a phase of synchronisation. This phase demands time, and depending on the time series length and complexity can be a cumbersome phase. The probability of the transitions can be used for the training. The second model, is further elaborated, and hence accepted by the researchers, where uncountable number of the studies were later performed based on this model, named Hidden Markov Model. This model considers both the priori and posteriori probabilities in its learning process. The third model, has recently received interests from the artificial intelligence community, especially after which several extensions of the model were integrated with the deep learning methods. This model, named recurrent neural networks, is being widely used in different fields of science and engineering, such as natural language processing, biomedical engineering, and media studies. This section will describe the three models in a general descriptive form, however, further elaborations of these models will be discussed in the other chapters where hybrid models, or deep recurrent neural networks will be explained in detail.

6.1 Dynamic Time Warping (Structural Classification)

In many applications, especially for the cases in which classification of the sequential data is speculated, we encounter with C classes of time series, each containing a number of N_c sample ($c = 1, ..., C$), and a testing time series x_i ($\forall i : x_i \in \mathrm{R}^{M,T}$) is attempted to be classified based on its similarity to the classes. The similarity is not limited to the contents of the input time series only, and variations of the time series are taken into account too. For a classification problem using a single vector, simple distance calculations such as Euclidean or Mahalanobis distance measurement can provide at least a measure, and in many cases do so with a classification method, to solve the problem. In time series classification, however, this is by far a more complicated problem. In addition to the stochastic behaviours of time series, synchronization of the time series, is a major point to consider. Initial points of a number of time series can be different while their contents are similar, and this offers a measure for similarity calculations which is rather complicated. This task looks to be even more complicated, when it comes with the multidimensional time series. In such the occasions, variation of time series, or more scientifically saying, "dynamic contents" of a testing time series should be compared to the time series for each class, and the testing time series is assigned to the class which gives the most similarity in terms of its dynamic contents. An important question raises up, here at this point, is: how to find a way to quantify similarity of dynamic contents for two different time series, perhaps with different length and different initial points? Dynamic time warping offers an algorithmic method to respond to this important question [116][103]. Application of this method is rather pronounced in the studying of sequential data with a relatively low level of uncertainty in its dynamic contents [66][19]. Nevertheless, dynamic time warping can often be an option to be studied in real world scenarios. Dynamic time warping takes the structure of the dynamic contents into account to find the similarity between two time series [138][111]. The classification method is very similar to the KNN method (see Section 3), with the difference in calculating the similarities. In KNN, the classification is performed according to the similarity of the samples of M dimensional vectors, likewise for dynamic time warping with the difference that the similarities are calculated for the time series of $M \times l$ dimension instead of the vectors [110][26][37]. This associates dynamic learning with the method. Figure 6.2 illustrates a typical block diagram of dynamic time warping classier:

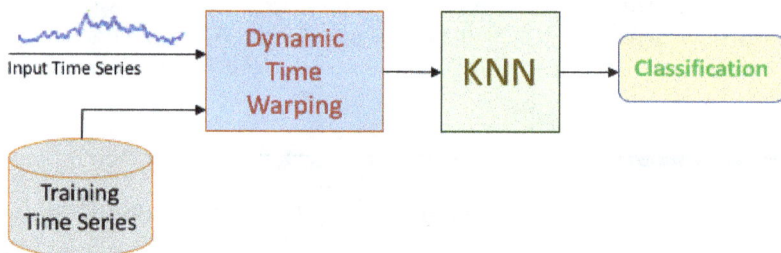

FIGURE 6.2: Block diagram of a classification using dynamic time warping.

Considering the two time series $X_1(t)$ and $X_2(t)$ with a different length of L_1 and L_2. The length of a time series, or in another word the number of the sequences in the time series, is given and the similarity of the two time series is to be calculated. Likewise the KNN where the similarities are calculated using distance measurement, typically Euclidean distance, dynamic time warping invokes the Euclidean distance technique for finding the similarity of the two time series. The distance of the time series is calculated as the global distance which is a sum of the distance at all the sequences. There are two important points to be considered when the global distance is calculated for the time series. The first point is the fact that the two time series can essentially have different lengths. The second point is less generalised, but can be seen in some cases, that the two time series may be asynchronous in the terms of their initial points. The dynamic time warping algorithm considers both the two points in its algorithmic structure for the distance calculation. To this end, the two time series are set in the horizontal and vertical of 2D coordination, named Time-Time matrix. The distance measurement in its simplest and quickest form of dynamic time warping is based on the linear alignment implementation, where the two time series are linearly aligned. An important assumption for linear alignment, governs to two time series to be synchronous in time, otherwise the calculated distance cannot give a realistic measure about dynamic contents of the two time series. Figure 6.3 shows a typical case of the linear alignment.

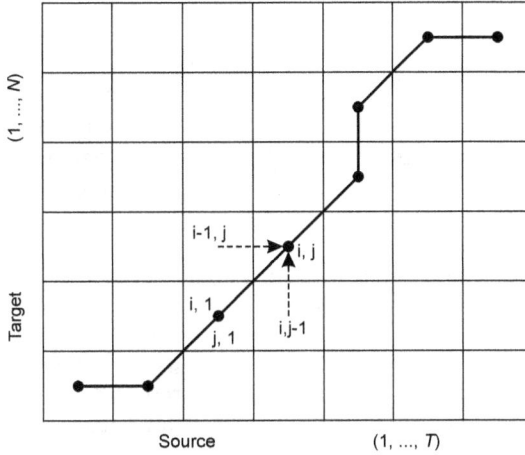

FIGURE 6.3: Linear alignment for calculating similarity of two time series using time warping method.

Similarity of two time series, as reflected by the distance of the time series, is fulfilled through the following algorithm for the linear alignment form of dynamic time warping:

Algorithm 5 Linearly-Aligned Time Warping

1: **procedure** LTW($\langle x_i \rangle, \langle r_j \rangle$))
2: Linearly align $\langle x_i \rangle$ and $\langle r_j \rangle$ $\qquad\qquad\qquad\qquad\qquad \triangleright i = 1, ..., N \;\; j = 1, ..., T$
3: $j \leftarrow round((1:T)N/T)$
4: **For i=1:T**
5:   $D(i) \leftarrow D(i-1) + d(x_i, r_j)$
6: **end for**
7: **return** $D(T)$
8: **end procedure**

In Algorithm 5, the $round(.)$ is the round function to the closer number, $D(i)$ and $d(x_i, r_j)$, is the accumulated and Euclidean distance, respectively.

Dynamic time warping with linear alignment, although it is simple and quick, but it can rarely offer an efficient measurement, and is mostly used for the purpose of initialisation for other methods. Therefore, a generalised form of dynamic time warping has been introduced in which an optimisation procedure is performed to align the two time series, asynchronously allocated in the Time-Time plane. This form of dynamic time warping which demands more calculation power, is often purposed for the cases where the two time series are not synchronised in terms of their onset points [35][65][5]. An optimisation process is performed in order to find the optimal match path. The process is performed iteratively through a pursuit for finding an optimal distance. As you will see in this section, in many advanced forms of dynamic time warping, this optimisation process is constrained with a number of the general, as well as the specific, heuristic or sometimes empirical, constrains for better learning. There are a set of the specific constrains, employed for the learning, intuitively presumed according to the prior information from the training data. This will be discussed later in this section. In any case, the general constrains must be necessarily fulfilled, otherwise the learning performance is not guaranteed. The general constraints are listed as follow:

- **Monotonicity:** The frames for calculating the distance should be incremental. A backward trace is not allowed.

- **Boundary condition:** The frames should be, started with the first sequence of the two time series, and terminated at the last sequence of the both time series. It cannot be started from any other frame, e.g., the second frame.

- **Continuity and symmetricity:** all the frames of the two time series must be used in finding the best matching path.

One should take this point into account (as was mentioned before) that all the time warping-based methods including dynamic time warping, rely on distance measurement between two time series (that can be multi-dimensional time series): an input time series and a time series from the training data (see Figure 6.2). The two time series can be asynchronous. The classification is performed based on the calculated distance with reference to all the samples of the training data. The distance measure does not play any role in the method generalisation. Assuming we have two **multidimensional** time series $X = [x_1, ..., x_N]$ and $R = [r_1, ..., r_T]$ (please pay attention to the point that the time series lengths can be different $N \neq T$), regardless of the distance calculation formula for the two vectors x_i and r_j, taken from X and R, the distance of the two vectors is denoted by: $d(x_i, r_j) = \|x_i - r_j\|$. In order to meet the continuity and symmetric conditions, the distance is allowed to be measured using one of the paths, depicted in Figure 6.4:

The horizontal path in parallel to the x axes is called deletion, while the vertical one along the y axes is named insertion. The diagonal path is termed by a match. The optimisation process for finding the best transition is based on calculating the accumulated distance, using the following derivation:

$$D(i, j) = min[\, D(i-1, j), D(i, j-1), D(i-1, j-1)\,] + d(x_i, r_j) \qquad (6.1)$$

where $D(i, j)$ denotes the accumulated distance, starting from the first vectors of X and R, and continuing until the last vector i and j from X and R, respectively (see the boundary condition). The $d(x_i, r_j)$ is named the local distance. The global distance is obtained by calculating all the accumulated distance for the last vectors of the two time series. Algorithm 6 illustrates the calculation of the distance for the two input time series.

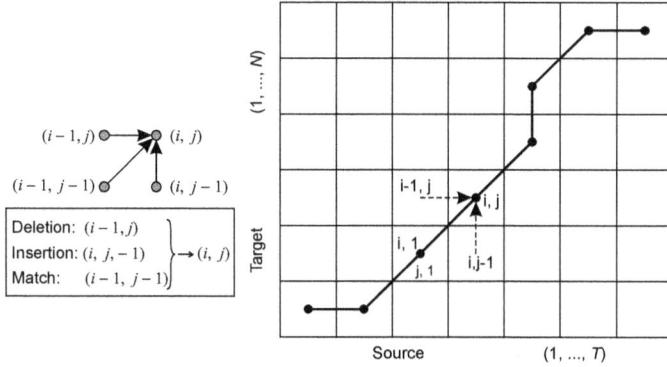

FIGURE 6.4: The path which are allowed in a way to fulfil the above-mentioned criteria.

Algorithm 6 Dynamic Time Warping

1: **procedure** DTW($\langle x_i \rangle, \langle r_j \rangle$))
2:    **For i=1:T**
3:       **For j=1:N**
4:          $D(i,j) \leftarrow min[\, D(i-1,j), D(i,j-1), D(i-1,j-1)] + d(x_i, r_j)$
5:       **end for**
6:    **end for**
7:    **return** $D(N,T)$
8: **end procedure**

In many applications, the performance of the method can be improved by assigning proper weights to the local distance, $d(x_i, r_j)$, in Eq. 6.1. This modification yields an improved learning, named weighted dynamic time warping, in which transition to the states charges a transition penalty λ. The transition penalty contributes in calculation of the accumulated distance. Multiple transition penalties can be defined to secure that the learning process converges to a valid result. Equation 6.2 shows the calculation of the weighted distance for one memory cell (see also Figure 6.5):

$$D(i,j) = D^* + \lambda^* d(x_i, r_j) \tag{6.2}$$

Use of the weighted dynamic time warping allows integration of prior knowledge which can bring an empirical risk to the classification method.

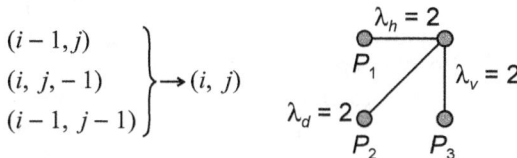

FIGURE 6.5: Assigning different weights to the warping paths.

As was previously mentioned, dynamic time warping only provides a means to measure the similarity of time series dynamic, and the ultimate classification is performed by another classifier such as KNN. In this case, the global distance must be normalised by the length of the two time series, in order to provide a fair metrics of similarity, when the time series length are different, and especially when the KNN is attempting to classify the patterns according to the calculated distances (the distance with all the samples).

$$\overline{D}(X, R) = \frac{D(N, T)}{N + T} \tag{6.3}$$

Without losing the generality, one can use several memory cells depending on the dynamic depth that is expected for the analysis. An example can be:

$$(\lambda_h, \lambda_d, \lambda_v) = (2, 1, 2) \tag{6.4}$$

Hence, a simple algorithm for implementing the weighted dynamic time warping is illustrated as follows:

Algorithm 7 Weighted Dynamic Time Warping

1: **procedure** WDTW($\langle x_i \rangle, \langle r_j \rangle$))
2:      For i=1:T
3:          For j=1:N
4:             $D^* \leftarrow d(x_i, r_j)$
5:             $D(i, j) \leftarrow min[\, D(i-1, j) + \lambda_h D^*, D(i, j-1) + \lambda_v D^*, D(i-1, j-1) + \lambda_d D^*\,]$
6:          end for
7:      end for
8:      $\overline{D}(X, R) \leftarrow \frac{D(N,T)}{N+T}$
9:      return $\overline{D}(X, R)$
10: **end procedure**

This is to some extent similar to the memory in the deep learning methods such as long and short term memory which will be explained in a separate chapter. Finding the learning weight can be based on the intuition, empirical or even by using hybrid methods where the weights are obtained by other dynamic classifiers. Sometimes the calculation path is restricted by the algorithms in a way not to permit certain distance measurements. This is performed relying on the fact that two well-aligned time series will likely result in a path which is not too far from the diagonal. Figure 6.6 illustrates a number of the possible constraints, used to restrain the alignment path.

Dynamic time warping demands small memories from the processor comparing to the other alternatives. However, as with many structural classification methods, a common problem of dynamic time warping which made it less appreciated compared to the other alternatives, is its relatively prolonged testing phase, especially when the training data size is large. For small size of the training data, dynamic time warping can be absolutely an option, as it

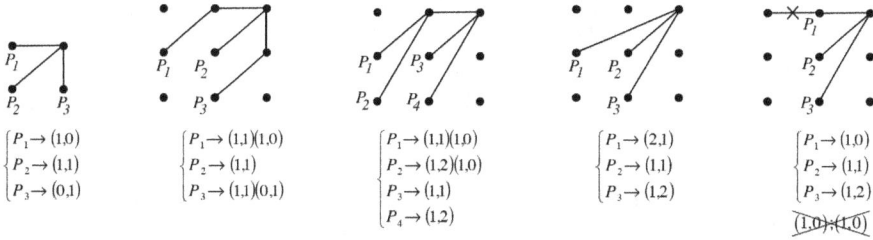

$$\begin{cases} P_1 \to (1,0) \\ P_2 \to (1,1) \\ P_3 \to (0,1) \end{cases}$$

$$\begin{cases} P_1 \to (1,1)(1,0) \\ P_2 \to (1,1) \\ P_3 \to (1,1)(0,1) \end{cases}$$

$$\begin{cases} P_1 \to (1,1)(1,0) \\ P_2 \to (1,2)(1,0) \\ P_3 \to (1,1) \\ P_4 \to (1,2) \end{cases}$$

$$\begin{cases} P_1 \to (2,1) \\ P_2 \to (1,1) \\ P_3 \to (1,2) \end{cases}$$

$$\begin{cases} P_1 \to (1,0) \\ P_2 \to (1,1) \\ P_3 \to (1,2) \end{cases}$$
$$\overline{(1,0)(1,0)}$$

FIGURE 6.6: Some examples of different possibilities to restrain the warping paths.

offers low structural risk, and the implementation detail is entirely interpretative. This is in contrast with the neural network-based methods. As a rule of thumb, this method is a perfect option for the post processing of dynamic data, resulting from applying stochastic based methods to an input time series, or even when there is a clear dynamic path of the sequential data which is expected for each class of the training time series.

6.2 Hidden Markov Model (Statistical Classification)

6.2.1 Model-based analysis

In many practical applications of time series analysis, we are faced with a set of the sequential data, alternatively called sequential patterns, created in an order which may look like a random sequence. The sequences can be generated by a certain system with a rationality with a certain extent of random behaviour, however, the ration of the system in generating the time series is neither perceived at the first glance, nor seen by looking at the time series itself. An observer sees a sequence of pattern, so called observation, but the system behind is hidden. This system is assumed to situate a set of the states, which are most likely to be hidden, and each state can generate a set of the patterns, profiled by their probabilities. The observer can see the pattern generated by the system, and attempts to identify the state of the system which produced the chain of observation.

The system model is the so called "hidden" to the observer, unlike the time series which is directly observed [32][109]. The main goal of many methods for time series analysis is to find the model from which the time series is generated, where the model is entirely hidden [91][79]. In most of the cases, a number of the models are predicted and the one with the utmost likelihood of creating the time series is recognised as the predicted model [113][148]. Classification of the time series will be easy, after recognising the model, based on the model parameters [25][80]. In fact, the system with the utmost likelihood to generate the observed time series, is assigned to the time series. This process is performed all through the training phase based on the probabilistic models and calculations [100][6][16]. In another word, in the test phase, an unknown time series is processed and likelihood of matching the time series with the predicted models is found individually for each model. The model with the utmost likelihood would be the class of the time series. This manner of time series analysis is called "model-based" analysis. Model based analysis can provide not only information about the time series class, but also about the state of the time series in terms of the model parameters.

6.2.2   Essentials of Hidden Markov Model (HMM)

Hidden Markov Model is one of the widely known model-based methods for time series analysis [113]. It provides an excellent tool for finding the model from the time series, especially when the time series resembles stochastic behaviours where other methods fail to provide efficient performance. There are different methods for using HMM in time series classification, considering time series as a set of the sequential data with random behaviour, but the hidden model behind might be deterministic, consisting of clearly known states. The art of HMM lies at this point that the state definition of the model (the possible states of the model named state diagram) resembles strict and deterministic, but creates random patterns with the known probabilities. The state-diagram of the model by itself must be presumed, but the model parameters are found by using any of the learning methods, which will be described in the following sequels. Before dealing with the model identification, a number of the definitions, used in the HMM model identification, are explained. A Markov model is fully described by having all of the following attributions:

1. State diagram model:

A state diagram model contains possible states of a hidden system, creating the observed sequential data/patterns. The state diagram also includes all the possible transitions between the states, based on the previous symbol. It is important to note that HMM is well-known as a probabilistic model, fully based on probability calculation, and the probabilities are all found relying on a set of the symbol. This is also the case for the state model as well as state transitions along with the state interconnections. To make the state diagram clarified, an example of a common healthcare problem is explained. A certain influenza disease manifests different states during the disease course including: fever, headache, running nose, and cough. These are accounted for by the state of the disease course. At each state, the three indicators composed of, the value of the White Blood Cell (WBC), the value of the Cretin Reactive Protein (CRP), and result of the Throat Bacterial Culture (TBC) are employed for diagnosing the type of influenza and also medication prescribed for the disease management. Progress of the disease might be different in person to person, but the general disease phases can include the following state diagram:

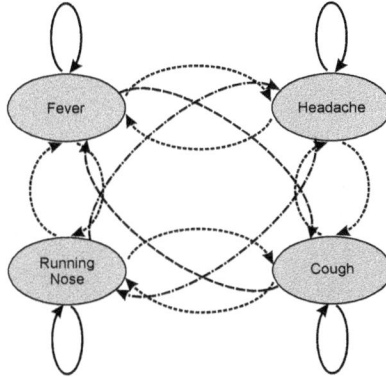

FIGURE 6.7: An example of the state transition diagram for an influenza disease.

The indicators are treated as the sequential data, as the observation shown in Figure 6.7, and the model creating the observation is required to be found.

Let's assume that the following observation is measured from a patient, and we have four different models of influenza, the objective is to find the influenza type from the sequence of the measured data. By using HMM methods, we can find the hidden model, or alternatively the disease type, creating the observation.

2. Symbols for the state model:

In the above-described example, there are three types of the indicators, two with the integer values including WBC and CRP, and one with positive or negative symbol which is TBC. It is necessary for a HMM to define the number of the symbols as well as the type of each symbol found in the symbols. For the rest of the section, it is assumed that there are $\mathcal{O}$ symbols of categorical data ($\in Z$) in the model.

3. State initial probabilities:

Probability of states of HMM at zero time is regarded as the state initial probabilities, denoted by π_i. In some of the applications this probability is considered to be identical for all the states.

4. State transition probabilities:

The conditional probability of moving from the state S_i to the state S_j is named posteriori probability for that state. The state transition probability is denoted by:

$$A = \{a_{ij} | a_{ij} = p(S_j | S_i), \ i = 1, ..., \mathcal{S}, \ j = 1, ..., \mathcal{S}\} \tag{6.5}$$

where $p(|)$ denotes conditional probability and $\mathcal{S}$ is the total number of the states. The state transition probabilities A, is a square matrix with $\mathcal{S}$ rows and $\mathcal{S}$ columns. The state probabilities are required for finding the model of a chain of the sequences produced by the Markov model.

5. Symbol probabilities:

Probability of a symbol O to be generated at the state S_i is named: symbol probabilities. The observation probabilities is a matrix with $\mathcal{O}$ rows and $\mathcal{S}$, denoting number of the symbols and states, respectively. The symbol probabilities are denoted as follows:

$$B = \{b_{jk} | b_{jk} = p(o_k | S_j), \quad k = 1, ..., \mathcal{O}, \ j = 1, ..., \mathcal{S}\} \tag{6.6}$$

One of the problems of HMM is that the state models together with the corresponding interconnections of the system which should be presumed based on the intuitions about the system which is supposed to be hidden. Understanding an appropriate model can be sometimes problematic.

6.2.3 Problem statement and implementation

HMM is commonly used for sequential data or time series analysis, either for the classification, or for prediction problems. Regardless of the application, one must always have prior knowledge about the possible states as well as the interconnection between the states before deciding to employ HMM for the analysis. This knowledge is intuitively obtained based on the model dynamics and nature of the system. Using this knowledge, there are three problems expressed by HMM, and often needed to be responded in all the applications:

1. Estimation:

In an estimation problem using HMM, a model $\lambda = (A, B, \pi)$ is already known, and probability of a given observation $p(X|\lambda)$, composed of a sequential data $X = \{x_1, x_2, ..., x_r\}$ is required to be estimated. In this case, the probability is calculated as the sum of the all possible paths, using forward recurrence algorithm, which is in turn based on three axioms:

a. The model should be situated to produce only one of the states, out of the total $\mathcal{S}$ states, at any given time t.

b. The output of the model is initialised by: $\forall S_j \ : \ \alpha_j = \pi_j . b_j(x_i)$

c. The algorithm is recursive which uses the following recursive formula:

$$\alpha_j(t) = [\sum_{i=1}^{\mathcal{S}} a_{ij}\alpha_i(t-1)] . b_j(x_i) \tag{6.7}$$

d. The algorithm is terminated when either a certain number of the iterations is performed, or a minimal change in the probabilities is observed, or even a combination of these two criteria is met and the probabilities are:

$$p(X|\lambda) = \sum_{i=1}^{S} \alpha_i(T) \tag{6.8}$$

2. Decoding:

In the decoding problem, a HMM model $\lambda = (A, B, \pi)$ is known and an observation of the sequential data is given $Q = \{q_1, q_2, ..., q_T\}$. An optimal sequence of the states is required to be found which can optimally justify the given observation $X = \{x_1, x_2, ..., x_T\}$ in terms of the probability. The term optimal in this definition implies on a states sequence which can most likely generate the observation sequence. Clearly an observation sequence can result from different state sequences and the one with the highest probability is the optimal state sequence. A well-known method for a decoding problem has been proposed by Andrew Viterbi, and accepted by many researchers [36]. In this algorithm, the probability is calculated for the most likely path only, in contrast with the original method in which the probabilities were calculated for all the paths. The Viterbi algorithm is a dynamic process programming, initialised by the following formula using the model parameters:

$$\delta_j = \pi_j.b_j(x_i) \quad \forall S_j \tag{6.9}$$

where the $\langle \delta_j \rangle$ is the sequence of the states. The algorithm is recursive with the following updating formula for the generated sequence of the states:

$$\delta_j = [\,max_i(\,a_{ij}.\delta_i(t-1))\,].b_j(x_i) \tag{6.10}$$

After a certain number of the recursions, or when the relative change of the decoded states is less than a certain threshold, the recursion is terminated at the following points:

$$p^* = max_i(\,\delta_i(T)) \tag{6.11}$$

3. Training:

In signal modelling using HMM, training implicates on finding an optimal model (state diagram in conjunction with $\lambda = (A, B, \pi)$), which is capable of accurately learning a set of the labelled observations. It is obvious that training in HMM is a supervised task. This leads to the problem of the following kind: the model state diagram is empirically presumed, but the manner of how to update the model parameters in a manner to optimise the probabilistic parameters, should be found. A widely used algorithm for training a Markov model, was proposed by Viterbi [36], which was later improved by different hybrid models resulted from the integration of multilayer neural networks with the model [149][82]. The hybrid models will be described in detail in the later chapter. Although Viterbi algorithm is time consuming and can lead to the risk of getting stuck in a local minima, it is worth to be described because of its favourable features such as simplicity.

Viterbi training algorithm is an iterative process, initialised by a linear alignment of the observations and the states according to the state probabilities, state transition probabilities, and symbol probabilities. Then, in the estimation process, the transition and the symbol probabilities are estimated by the following calculations:

$$\widehat{a}_{ij} = \frac{Number\ of\ transitions\ from\ state\ S_i\ to\ state\ S_j}{Number\ of\ transitions\ from\ state\ S_i} = \frac{N_{ij}}{N_i}$$

$$\widehat{b}_{jk} = \frac{Number\ of\ observations\ of\ symbol\ V_k\ at\ state\ S_j}{Number\ of\ observations\ at\ state\ S_j} = \frac{N_{jk}}{N_j} \quad (6.12)$$

At the third step, based on the estimated parameters, a decode process is performed and an observation of the states is created. This observation is aligned with the training data and the error is calculated. The same procedure is repeated until reaching stable values for the error, or a criterion of not exceeding a certain number of the iteration is met.

6.2.4 Time series analysis and HMM

It was shown that a HMM process is capable to up a model using observations of sequential symbols, or sequential patterns, as its inputs. A time series can obviously receive any real-valued random variable, hence a pre-processing level is sometimes required to turn the time series into a sequence of patterns, which will constitute the observations to be employed by the HMM. Figure 6.8 illustrates the process.

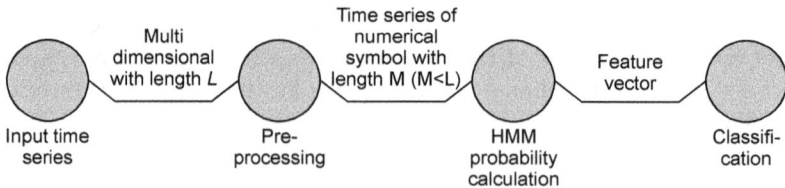

FIGURE 6.8: The process needed for classification using hidden Markov model.

This mapping from the real-valued time series, in general a multidimensional time series, to a sequence of the observations composed of patterns, is named pre-processing. This pre-processing often invokes a quantification technique such as distance-based quantification (Euclidean distant), KNN, Gaussian mixture model, or even MLP neural network, as the mathematical tool to perform mapping. In most of the cases a supervised mapping is preferred, nevertheless, there is no limitation to use unsupervised clustering methods, i.e., K-Means clustering to perform this mapping. Therefore the combination of the unsupervised and the supervised model-based method can be of special interest when we have multiple recordings of time series and would like to extract the patterns of information from the noise [40]. Nevertheless, such the models are mainly heuristic and can be faced with difficulties to be generalised, and therefore were not included in this book.

6.3 Recurrent Neural Network

Recurrent neural network is another form of the dynamic methods for learning the variant contents of time series and sometimes the sequential data. Architecture of a recurrent neural network contains one/several feedback of the outputs, which are used during the programming, both for the training and the testing. Figure 6.9 illustrates an architecture of the recurrent neural network.

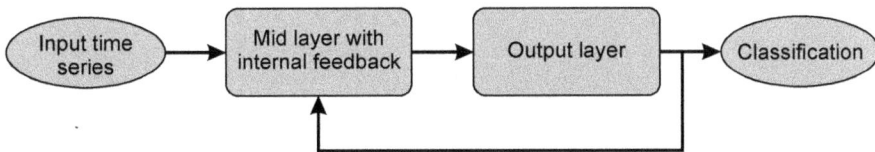

FIGURE 6.9: Block diagram of a recurrent neural network.

Recurrent neural networks offer the possibility of incorporating previous states of data, and also the classification result, into the learning process, through a predefined architecture of including the past content of the time series in the learning architecture. Such the past contents of time series are stored in the memory of the method, which differentiates the dynamic learning from the static learning such as MLP. Bear in mind that system architectures like the Time-Delayed Neural Network, even though introduce the potential to bring dynamic contents of the time series into the learning process, still are incapable to invoke the past contents of the system for decision making. As a matter of fact, these architectures do not employ dynamic programming, and the dynamic contents of the time series are treated as the input features, unlike recurrent neural network in which system outputs from the previous time instances are employed by the network for the classification. One should be aware that recurrent neural network is different from the adaptive method in this aspect that a recurrent neural network is trained only once, and when trained the learning weights are employed for the classification. Some of the learning weights are assigned to the feedback outputs at the previous time instances. In contrary, an adaptive method cannot take any decision and the parameters of the method are updated at certain time intervals.

In analogy to HMM, the method memory for a recurrent neural network acts as the previous symbol of a HMM, which participates in predicting the future of the method. Likewise the HMM in which deep understanding about the classification goal as well as the input data provided intuitions for designing state diagram in conjunction with the corresponding connections, a deep insight into the nature of the input data is necessitated when architecture of a recurrent neural network is going to be designed. Figure 6.10 shows architecture of a recurrent neural network, known as the Elman neural network, which is commonly employed for the classification purposes [34][146].

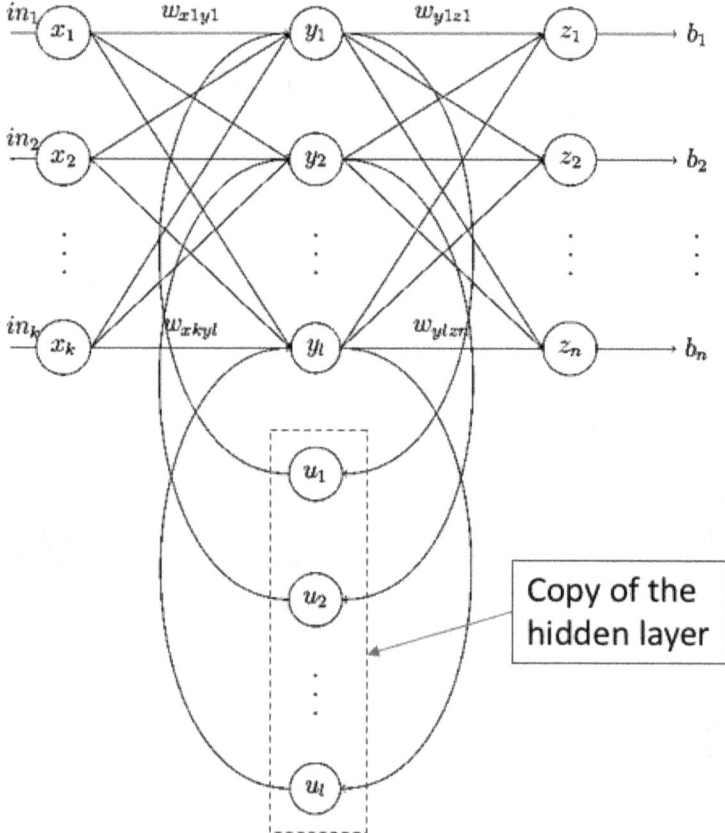

FIGURE 6.10: Block diagram of an Elman recurrent neural network.

As seen in Figure 6.10, the architecture is designed according to the applications and demands, which is also true for the training strategy. In this architecture, a copy of the hidden layer from the previous time sample is incorporated into the learning process. Another alternative to Elman neural network is known as Jordan neural network, in which the copy is taken from the output layer at the previous sample time [146]. In many cases of recurrent neural network, back propagation error method is invoked for the training, where the hyper parameters are obtained by an appropriate optimisation method such as natural inspired methods as defined by genetic algorithm and evolutionary computing.

Recently, recurrent neural networks have been profoundly elaborated by different architectures, each offering new elaborations compared to the previous ones. Long and Short Term Memory (LSTM) is one of the most commonly used architecture of recurrent neural network, whose application was broadly studied in different topics. In this book, recurrent neural networks will be explained separately in Chapter 11, where details and features of the recurrent neural networks will be scrutinised. This brief explanation was introduced in order to describe the most commonly used alternatives of dynamic programming for processing sequential data and also for time series analysis.

Part III

Deep Learning Approaches to Time Series Classification

7

Clustering for Learning at Deep Level

The concept of deep learning is focused on neural networks with many layers and neurons, even up to thousands, containing several hyperparameters for the training, needing heuristic and sometimes naturally inspired methods for the optimization. This can be certainly true to a certain extent, but definitely cannot cover the entire scope of deep learning. This chapter provides another view towards deep learning based on using unsupervised methods to help the learning process for extracting subtle contents of time series at the deep level. This chapter focuses on the potentials of the well-known clustering methods in the deep learning perspective, that can strengthen the learning process.

Segregation of data into a certain number of the groups, where the number of the groups may not be necessarily known in advance, in an unsupervised manner, i.e., clustering [102]. The number of the groups is assumed to be known in majority of the clustering algorithms, however, it is not necessarily a fixed number [118][147]. In some of the unsupervised methods, an optimal number of the groups is also found through the learning process. Regardless of knowing the number of the groups, a clustering method attempts to find data similarities and group the data according to the similarities quantified by a reckoning learning rule. This aspect of clustering has been sought in the modern deep learning to extract latent information from the deep level of the learning architecture.

Clustering from the beginning was deemed only as a method to extract similarities when we didn't have access to the real class of the data. Application of clustering as a tool was evolved in time for extracting the pure information out of the noisy data, by assuming that the variations appearing in the data contents come from the different sources of noise. To some extent, this unwitting contaminating source can be excluded from the data by using a suitable clustering method. An important assumption to such the applications is that the noise level is far lower than the data. It is certainly accepted that a number of the data points can be affected because of noise, but the general data population is expected to be discriminant even at the presence of various sources of noise. It is obvious that the discrimination power of a clustering method in categorizing the data according their similarities plays an important role, which can overcome the noise-caused data discrepancy.

The capability of a clustering method in extracting similarities within a group of data with an identical class, and also dissimilarity of data groups for different classes, can provide a powerful means for any kind of these learning methods. This ability is intuitively provided by different learning methods such as multi layer perceptron neural networks, however, many practical solutions were still lacking from a desirable performance. Even after development of the sophisticated neural network-based methods, still performance of the classifier is sometimes unsatisfying. The researchers have thereafter become motivated to enhance the learning capability of neural networks by increasing the learning layers of neural networks, that eventually led to innovative architectures, named deep learning methods. At this level, an important point that needs further attention, is the fact that blindly invoking deep learning to develop a classifier, can result in a high structural risk and unreliable performance of the classifier due the very large number of the hyper parameters associated with the deep learning methods. Nowadays, efficient design tools such as TensorFlow have made a breakthrough in designing the architecture of deep learning. However, using a simple and easy task, jeopardizes performance of these methods in terms of the structural risks. Nonetheless, powerful learning exists in the latent space of deep learning methods that elaborates the learning performance. In fact, art of the deep learning method lies in offering an integrated model in which the feature extraction and pre-processing levels, are incorporated into an integrated architecture. This is certainly an elaboration for the method at the expense of association high structural risk with the architecture. It is commonly seen in most of the neural network-based methods with many layers, that can exceed even up to hundreds of the layers.

Another common problem often seen is the blind use of many heavy architectures of deep learning methods, is the overfitting problem, caused by a large number of the hyper parameters in relation to the training data. This is an additional turning point that tends to increase structural risk of the deep learning methods. Such problems have been well resolved by the recent architectures of interpretive and understandable structures, introduced to the community of artificial intelligence. In these structures, an input layer of processing performs the learning at the deep level, either for the feature extraction, or for learning deep contents of the data. The input layer delivers the purified, or in another language quantified data, to the outer layers which performs the ultimate classification. In analogy to HMM, the input layer of a deep learning method performs a pre-processing as was explained in a separate block of a HMM-based method. The level of processing is, however, integrated with the rest of the architecture of the learning method. This fashion of the classification, which is sometimes named multi-scale classification method, offers an advantage to the HMM-based method, resulting from the fact that the learning of the deep level is not completely separated from the rest of the learning, unlike the HMM in which the pre-processing phase is completely independent, and separately trained. Therefore the integrated training can introduce an elaboration in learning process, resulting from an improved possibility of extracting similarities and discriminative contents of the data. As you will see in this chapter, clustering methods can be wisely invoked to extract subtle and very detailed contents of the information from the data, down at the deep level of the learning process.

This chapter deals with the time series analysis and the theoretical foundation is based on stochastic time series which can be contaminated by various sources of noise. Of course generality of the theories is not lost by using time series analysis as the case study, however, the main focus of this book is time series analysis. In this perspective, the reset of the methods will be focused on learning stochastic time series for the purpose of the classification, where the between-class similarities makes conventional classification methods incapable of learning discriminative details of the time series. Application of this methods can be clearly seen in time series of the physiological signals, which is a typical example of the signals with high complexity coming from the complicated source of the data. This application is today a central part of the context of biomedical engineering. This processing can be oriented towards time series classification either for the diagnosis, or for patient monitoring purposes. Some of the examples of this application will be briefly addressed throughout the sequels. An experience of two decades of studying biological signals for the diagnosis and monitoring, has taught the authors that not any fashioned method can lead the users to arrive at an efficient system by which a reliable performance can be achieved. In fact, benefits of many deep learning methods are limited for only very large scale companies who have access to a huge datasets of great data variety, to support an efficient training without been trapped in difficulties such as overfitting. This convinces the author to present the below methods by which efficient performance has been observed [61][63][45][60][57][59]. Another practical factor to be considered is complexity of the learning methods. In some of the applications of Internet of Things, a heavy deep learning method cannot be implemented at the edge level, especially when it comes with the federated learning. In addition to the memory needs for the training and testing, power consumption takes an important role in the internet of things. Nevertheless, a number of the sophisticated deep learning methods are designed to be implemented in such structures [52][152][58].

7.1   Clustering as a Tool for Deep Learning

Community of artificial intelligence, in its older fashion, described clustering as an unsupervised method where the data labels do not play any role in the learning process. This traditional description can no longer maintain its meaning, in some of the cases, for instance in the cases of deep learning when it comes to the learning of time series at the deep level. In this case, the learning process is assumed to be unsupervised, however, a clustering method is still invoked to assign a set of the labels to the data, according to the within-class similarities, and the between-class dissimilarities. Next, the labels are compared to their actual class, and the classification error is calculated. The feature set which provides the optimal performance (in most of the case the largest classification rate) is selected for the rest of the learning process. Figure 7.1 illustrates the way of using of a clustering method for learning time series contents at the deep level. In this manner of learning, clustering finds a feature set among a large number of the input features, which provides optimal discrimination power.

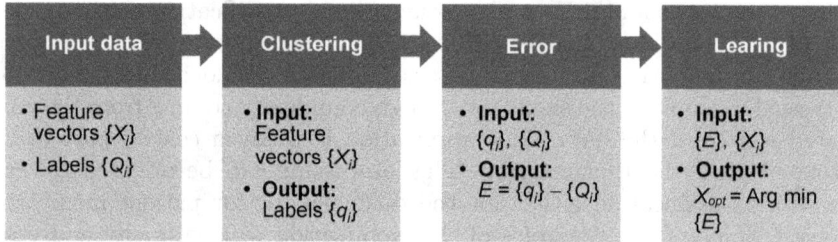

FIGURE 7.1: Clustering is used to find an optimal feature set at the deep level of the classification.

This structure comes from a multi scale classification, where subtle details of the time series are extracted by using a clustering method, and the resulting abstract of the feature vectors are employed by another dynamic learning method such as recurrent neural network or HMM. Let's assume that a clustering method assigns a label to an input data vector of dimension M, by the function $\phi : \mathbb{R}^M \to \mathbb{Z}$ defined by:

$$q_i = \phi(x_i), \quad \forall i : x_i \in \mathbb{R}^M, \quad i = 1, ..., N \tag{7.1}$$

Assuming the actual data labels to be $Q_i \in \mathbb{Z}$, the classification rate is calculated for an input dataset of size N as follows:

$$I = \frac{\sum_{i=1}^{N} \delta(q_i, Q_i)}{N} \tag{7.2}$$

where

$$\delta(x, y) = \begin{cases} 1, & \text{if } x = y \\ 0, & \text{if } x \neq y \end{cases} \tag{7.3}$$

It is usually difficult to characterise a time series, directly by its temporal contents, especially for the stochastic cases, unless the groups are quite discriminant in terms of the temporal variations. This task requires another step of processing to find an appropriate mapping (alternatively saying mathematical transformation) for the time series to another domain, or sometimes another space, in which different classes are better segregated. That's why mathematical transformation has become an important topic of machine learning. One of the most widely used transformation techniques is Fourier transformation. This powerful method, which has been profoundly used over many decades, maps temporal variation of a time series to a domain of complex numbers, named frequency domain, where the temporal variations can be characterised easier than using merely temporal contents of the time series. For the time series $x(n)$ of length N, the Fourier transform is calculated as:

$$X(\omega) = \frac{1}{N} \sum_{n=0}^{N-1} x(n)e^{j\omega t} \tag{7.4}$$

For stochastic time series, the Fourier transform is calculated over a windowed time series, and the square power of the complex numbers resulted from the Fourier transform for each window, is averaged over the windows. The result is indeed an estimation of the spectral contents of the time series which is affected by several parameters such as the window's length, the window's shape, and the window's overlap. This way of direct form of spectral estimation is named as "Periodogram" or simply "direct form of spectral estimation". This point is put into a better perspective in the followings.

For a temporal window w of length N_0, the windowed Fourier transform is calculated as:

$$X(\omega, t) = \frac{1}{N_0} \sum_{n=0}^{N-1} x(n+t)w(n)e^{j\omega t}, \quad t = [0, ..., N - N_0], \quad N_0 << N \tag{7.5}$$

where square power of the Fourier transform of a time series is known as spectral energy of the time series. For stochastic time series, windowing is always known as an important consideration to be "taken with a little bit of salt". Length of the windows is fixed and also the time series, in its full-length, must be covered by a set of the windows. Sometimes overlapping windows are employed to improve the time frequency resolution. For a better understanding, one should consider that choosing a long window yields losing the temporal resolution (see Eq. 7.4 in which t spans a short range since N_0 approaches N). On the other hand, too short window results in a low spectral resolution (N_0 in Eq. 7.5 becomes small). The trade-off between time and frequency resolution requires an appropriate selection of the window size. The readers can refer to the literature in discrete time signal processing to find more details about the trade-off between time and frequency resolution [104][132][32]. Dynamic contents of a time series can be found by using spectral contents of a temporal window sliding over the time series as defined by periodogram. Thus equation (7.5) becomes:

$$X(f, t) = \frac{1}{N_0} \left| \sum_{n=0}^{N-1} x(n+t)w(n)e^{j2\pi fn} \right|^2 \tag{7.6}$$

The window's shape is selected in a way to reduce side lobe leakage. More details are found in [104][132][32]. Here, the question of time series characterisation turns into another form: what frequency bands can be used to calculate spectral energies such that similarity and dissimilarity between different classes are optimally quantitated? This directly points to a learning process, remembered from the introduction where learning was introduced as finding similarities and dissimilarities between different classes. For stochastic time series, using only one level of learning may be insufficient, and therefore multi-scale learning methods, which were later recognised as deep learning are administered in most of the cases. The structure of such learning methods involves learning details of the time series at the deep level. Clustering methods can be well employed at this level of learning with some considerations. To this end, a set of the feature vector of length M is extracted from each temporal window, over the frequency bands B defined as follows:

$$B \doteq \{[f_1, f_2]^T : f_1, f_2 \in Z^+ \wedge f_1 < f_2 < F_{max}\} \qquad (7.7)$$

where F_{max} is the maximum frequency presumed for the analysis, that is often defined by the signal collection system. An example of such the restrictions in the real-world scenarios is the sampling frequency. In many practical situations, the maximum frequency is half of the sampling rate. Nonetheless, using the above definition for frequency band, can help the formulation of finding a set of the frequency bands whose spectral energies provide optimal discrimination power. Spectral energies calculated over of a set of frequency band, known as the sub-band energy, becomes:

$$P(b_k) = \frac{1}{T_0} \sum_{t=0}^{T_0-1} \sum_{b_k \in B} X(f, t) \qquad (7.8)$$

where T_0 is the total number of the temporal windows. One may refer to Eq. (7.5) and conclude that the total number of the temporal windows is $N - N_0 + 1$. This conclusion is not incorrect, but not generalised. This is only true if the temporal shift is one sample which is not favoured in terms of the calculations power, as it is a cumbersome process because of involving a high number of the temporal windows. In most of the practical situations, in order to make the process less demanding, the temporal shift is considered to be much larger than one sample. A percentage of the windows length, expressed as in percent, is mostly used, e.g., 75%, which causes a certain overlap over for two successive windows. This manner of using the overlapping windows is very common in most of the signal processing applications. In fact, use of the temporal shift of one sample, might be seldom seen specially when the signal length is large.

Finding such a set of frequency bands is indeed a kind of the learning process at the deep level. This way of learning involves a pursuit over the spectral energies calculated as defined by Eq. (7.6) using all the combinations of different frequency bands and selecting the set which provides an optimal value of discrimination power. This set of the frequency bands is named discriminative frequency bands. As such the pursuit can be performed in different ways. Hill-Climbing algorithm is one of the alternatives to this end. Assuming $\mathcal{I}$ is the value of the discriminative function resulted from a set of the frequency bands with the utmost discrimination power, these discriminative frequency bands can be found through the following algorithm:

Algorithm 8 Learning Discriminative K Frequency Bands using Hill-Climbing

---

1: **procedure** DFB($\langle X_i \rangle, K$))
2: Calculate $X_i(f, t)$ ▷ Eq. 7.6
3: Calculate $P(b(X_i(f, t)))$ ▷ Eq. 7.8
4: $I_1 \leftarrow \arg\max_m \mathcal{I}(P(b(X_i(f, t))))$
5: **For j=2:K**
6:      $I_j \leftarrow \arg\max_m \mathcal{I}(\langle[P(I_1), ..., P(I_{j-1}), P(b(X_i(f, t)))]\rangle)$
7: **end for**
8: **return** $[I_1, ..., I_K]$
9: **end procedure**

Clearly, this is an iterative and time demanding algorithm. The number of the pursuits required to find K discriminative frequency bands depends on the total number of the frequency bands that exist in (7.7). In order to find this number, we begin with the assumption of $K = 1$ to find the number of the pursuit for one discriminative frequency band. In this case, we firstly assume that f_1 in (7.7) is set to 0. In this case there are F_{max} frequency bands to be examined. Next, the f_1 is set to 1, and obviously the number of the frequency bands is $F_{max} - 1$. This will continue until the last, where ($f_1 = F_{max} - 1$) and there is only one frequency band. Hence, the number of the pursuit is found by accumulating the number of the bands at each iteration:

$$\sum_{b \in B} b = F_{max} + (F_{max} - 1) + ... + 1 = \frac{F_{max}(F_{max} - 1)}{2} \tag{7.9}$$

One can easily see that using the above-mentioned Hill-Climbing method yields the total number of the pursuit for finding K discriminative frequency band by the following equation:

$$Pursuits = \sum_{j=0}^{K-1} \left(\frac{F_{max}}{F_{max} + 1} - j \right) \tag{7.10}$$

This algorithm demands a time-consuming process, even though it is efficient. Other optimisation methods such as evolutionary computation or genetic algorithm can be invoked instead of using the hill-climbing method in Algorithm 8.

It is again stated that any clustering method can be employed for finding a set of the frequency bands which provides an optimal discrimination. This can be assumed as learning at the deep level in which an optimal feature set is learned, that is indeed a way to integrate the feature extraction with the architecture of the learning process, which can serve as a learning method at the deep level, as you will see in the next chapters. Depending on the nature of the clustering method chosen for the learning purpose, certain attention must be paid. Here the clustering method is not invoked for labelling, instead is employed as a criterion for quantitating the discrimination power of the chosen method. Such the elaborating aspect of clustering method takes nature of the method into consideration rather than the labels assigned by the method. A number of the well-known clustering methods are formulated in the following sequels for this learning purpose.

7.2 Modified K-Means Method

K-Means has been previously described in Chapter 3, where it was shown that this method segregates the clusters of data according to the statistical distribution, reflected by the mean of the centroids. One of the problems of K-Means in attaining a stable clustering is initialisation of the iterative method for the clustering. In fact, sometimes the clustering might land in different clustering results because of the different initialisation data. Nevertheless, this method is invoked for learning the deep contents of the data according to the classification rate obtained for each feature set. In order to prevent different clustering results in each trial of K-Means, a fixed initialisation process is employed, in which the absolute value of the feature vectors is calculated and uniformly distributed over the classes for the initialisation. The algorithm starts with one feature of spectral energy, and equally distributed based on their values. Then, the features of spectral energies are added one by one. At each step of adding a new feature, K-Means algorithm is applied to the data to make a new clustering. Let's assume the frequency bands, defined in (7.7), are employed to form a feature vector as in (7.8). Then, augmented spectral energy for a set of the frequency bands is calculated as in Algorithm 8 and the classification rate of the resulting feature is calculated and the classification error with respect to the actual classes is found for each frequency band. The band with minimal classification error is selected as the discriminative band. In fact, the discriminative function of Algorithm 8 becomes the classification error resulted from the K-Means clustering.

7.3 Modified Fuzzy C-Means

The previous method is based on grouping the data samples to a pre-defined number of the clusters according to their relative distance. Then, the centroid of the clusters are calculated and iteratively updated until reaching a stable segregation. At each iteration, a single data sample belongs to only one cluster, even though the cluster labels may be variant over the iterations. The c-means, follows a similar clustering method, with the difference that a single data sample may belong to all the clusters, but with a different membership value. This is another parameter associated with the clustering method, seen in the fuzzy-based methods including c-means, describing how a parameter can be part of a cluster. Membership value of a data sample to a class is indeed the degree of belongingness of the data sample to the class. Membership is different from probability, in this sense that the sum of the membership values of a single sample data over the whole available clusters can be unequal to one, in contrast with the probability which always yields a unity sum over the classes. Membership is a broader definition than probability since this summation can be even one, depending on the developer's design. Assuming that there are N data samples $X = [x_1, ..., x_N]$, which are tried to be segregated to C clusters, and the feature vectors are defined in Eq. 7.8. In C-Means clustering method, an initial value is guessed for the centre of each cluster. This initial guess together with the memberships are updated through an iterative procedure, where an objective function plays an important role in the updating procedure. The objective function, Ob, defined as follows, is attempted to be maximised in the iterative procedure:

$$Ob = \arg_C \min \sum_{i=1}^{N} \sum_{j=1}^{C} w_{ij}^m \|x_i - c_j\|^2$$

$$w_{ij} = \frac{1}{\sum_{k=1}^{C} \left(\frac{\|x_i - c_j\|}{\|x_i - c_k\|} \right)^{\frac{2}{m-1}}} \tag{7.11}$$

where m ($m \in (1, \infty)$) is the degree of fuzziness. A higher value for m corresponds to a higher fuzziness. Likewise to the K-Means algorithm, an iterative algorithm can be invoked for the learning at the deep level using clustering error as the cost function. The label to the data samples are assigned according to the membership values to the clusters. The membership value is found by:

$$c_k = \frac{\sum_x w_k(x) m x}{\sum_x w_k(x)^m} \tag{7.12}$$

This cost function is contributed in the learning process at the deep level, by which an optimal feature sets are found. Although C-Mean and K-Means clustering methods look to be very similar to each other, the accuracy resulting from these methods can be very different in many applications of deep learning.

7.4   Discriminant Analysis

In the previous method, time series contents were learned at the deep level, using K-Means clustering as the mathematical unsupervised mapping method and the resulting labels were compared to the actual labels where the classification rate is employed as a metric for quantification of the discrimination power, and consequently the learning process. The pursuit was repeated towards finding an optimal value for the metric. Without losing generality of the method, one can employ other statistical methods instead of K-Means clustering to proceed the learning process at the deep level. Discriminant analysis method can be alternatively employed for such the quantitation, either by itself or in a modified form by the transformed version obtained using the eigen vectors. Assuming a training dataset of size N, contains C groups, each containing N_i $(i = 1, ..., C)$ number of the time series. Spectral contents of each single time series is obtained from the Eq. 7.8. One way to quantify capability of data segregation, is to use the technique presented as the discriminant analysis and Fisher criterion. Recalling from Chapter 3, the Fisher value is calculated by:

$$Fisher \ \ Value = \frac{Between \ \ Scatter}{Within \ \ Scatter}$$

$$S_B = \sum_{i=1}^{C} p(\omega_i)(\mu_i - \mu) \cdot (\mu_i - \mu)^T$$

$$S_W = \sum_{i=1}^{C} p(\omega_i) \cdot \Sigma_i$$

$$\Delta = \frac{|S_B|}{|S_W|} = \frac{|\sum_{i=1}^{C} p(\omega_i)(\mu_i - \mu) \cdot (\mu_i - \mu)^T|}{|\sum_{i=1}^{C} p(\omega_i) \cdot \Sigma_i|}$$ (7.13)

$$\mu_i = E\{x|\omega_i\}, \quad \mu = E\{x\} = \sum_{i=1}^{C} p(\omega_i) \cdot \mu_i$$

$$\Sigma_i = E\{(x - \mu_i) \cdot (x - \mu_i)^T |\omega_i\}$$

where $p(\omega_i)$ is the probability density function of the random variable ω_i and Δ is the Fisher value.

For the spectral feature vectors, $P_i = [P_{1,i}, ..., P_{M,i}]$ $(i = 1, ..., N)$, as defined by Eq. 7.8, the learning process begins with one feature with optimal discrimination power, and then continues by adding the other the sub-optimal spectral features one by one until reaching the last one. In order to find the first discriminative frequency band, b_1, one dimensional Fisher value is employed to facilitate the process:

$$\Delta = \frac{\sum_{i=1}^{C}(\mu_i - \mu)^2}{\sum_{i=1}^{C}\sigma_i}$$

$$\mu_i = \frac{\sum_{j=1}^{N}P_{1,i}\delta(i,j)}{N_i}$$

$$\mu_i = \frac{\sum_{j=1}^{N}P_{1,i}}{N} \tag{7.14}$$

$$\sigma_i = \frac{\sum_{j=1}^{N}\left((P_{1,i} - \mu_i)\delta(i,j)\right)^2}{N_i}$$

$$\delta(x,y) = \begin{cases} 1, & \text{if } x = y \\ 0, & \text{if } x \neq y \end{cases}$$

The following algorithm shows the learning process at the deep level using Fisher method:

Algorithm 9 Learning Discriminative K Frequency Bands using Fisher criteria

---

1: **procedure** DFB($\langle X_i \rangle, K$))
2:    Calculate $X_i(f, t)$ ▷ Eq. 7.6
3:    $P_1 \leftarrow P(b(X_i(f, t)))$ ▷ Eq. 7.8
4:    Calculate $\langle \mathcal{I}_i(P_1) \rangle$ ▷ Eq. 7.14
5:    $b_1 \leftarrow \arg\max_m \langle \mathcal{I}_i(P_1) \rangle$
6:    **For j=2:K**
7:        $P_{j,i} \leftarrow P(b(X_i(f, t)))$ ▷ Eq. 7.8
8:        $I_i \leftarrow \langle \mathcal{I}([P(b_{1,i}), ..., P(b_{j-1,i}), P_{j,i}]) \rangle$ ▷ Eq. 7.13
9:        $b_j \leftarrow \arg\max_m I_i$
10:    **end for**
11:    **return** $[b_1, ..., b_K]$
12: **end procedure**

The optimal bands are added to the feature vector, one by one towards achieving the feature sets of length K providing optimal segregation. It is of critical importance to consider that for a certain datasets, the determinant of the S_B and S_W matrixes in (7.13) may tend to a close vicinity to zero, which put calculation of the Fisher value into the risk of overflow, or in another term leads to singularity and incorrect learning. The algorithm therefore, sets a criterion of the bounded determinant meaning that the feature sets, creating an extraordinary high value of the determinant of the two matrices. The bands with a singular value, or alternatively a low value of the determinant is excluded from learning. This iterative algorithm can be boosted by involving eigen vectors of the Fisher matrix, using the following derivations:

$$\Delta = \frac{\left|W^T S_B W\right|}{\left|W^T S_W W\right|} \tag{7.15}$$

where $W = [w_1, ..., w_M]$ corresponds to the matrix resulting from M eigen vectors with the largest eigen values. The only difference comes to the implementation of the discriminant analysis, where in Algorithm 9 the learning process is quantified by the eigen value of the Fisher matrix in Eq. 7.15 rather than merely the Fisher value. In any case, the determinant of the Fisher matrix must be checked to avoid singularity. The rest of the algorithm remains similar to the previous ones. Using the eigen vectors of the Fisher value brings rather secure learning to the process against the background noise. This comes from the fact that the principal components correspond to the eigen values project the background noise. Thus, to some extent, the background noise can be eliminated by excluding the components with low eigen value.

7.5   Cluster-Based vs Discriminant Analysis Methods

The cluster-based learning methods, including K-Means, C-Means or any other clustering method, all work relying on the classification rate. Although methods like K-Means take the statistical distribution of the data into consideration in their clustering process, the final decision for selecting the discriminative optimal feature sets are made merely based on the classification rate. Regardless of the distribution of the data, the classification performance undertakes the learning process. This can be flawed for the small or medium size of the training data, as performance of the method is potentially at the risk of the instability when it comes with a large testing data out of the training data. To put this point into a better perspective, consider Figure 7.2, depicting distribution of 2D data for two cases.

In Case 2, the classes are well segregated in terms of the classification rate, however, the data is distributed in a way that a large number of the borderline data samples are seen around the border. Therefore, for a large dataset out of the training data, the probability of mixing the data samples of the two groups is something to be considered. In the other graph of the figure, the Case 1, the two classes are better segregated compared to the previous one, even though the classification rate might be lower than the other one, because of some outliers from the two classes interlacing the other class, and therefore are wrongly classified. One can easily have an intuition that for a large dataset out of the training data, a high probability of repeating more or less a similar result is foreseen. Fisher value on the other hand, takes distribution of the data samples into account, by using the joint "within scatter" and "between scatter", simultaneously. The two graphs depicted in Figure 7.2

Case 1   Case 2

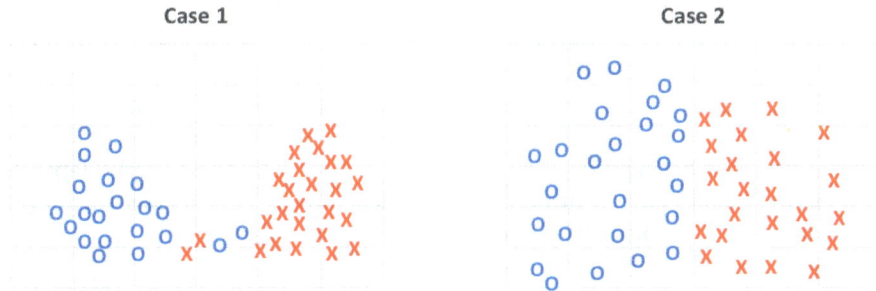

FIGURE 7.2: Two cases of data distribution: One with a higher Fisher value (Case 1) against the other one with a better classification rate (Case 2) but poor Fisher value. A combined form of the two criteria, the Fisher value and the classification rate, can result in a better performance.

are exaggeratedly demonstrated in order to illustratively show performance of the two approaches of learning at the deep level: the cluster-based and discriminant analysis methods. Obviously such the extreme cases are unrealistic to be observed in real-world scenarios.

7.6 Combined Methods

The supervised clustering methods, prescribed for learning at the deep level in the previous sections, all employ spectral contents of the time series over a short length window to learn dynamic contents of time series at the deep level. As you will see in the coming chapters, another level of the learning is necessitated to extract dynamic contents of the time series over the surface constituted of the learning outcomes from the deep level. Therefore a multistage structure demands a smaller size of the parameters for deep learning and is suitable for small and medium size training data. In this situation, a problem that can commonly arise for the clustering-based method is that, one can reach to several frequency bands all resulting with a similar classification error. Assuming that there are K frequency bands all yielding an identical classification error, an important question is: which frequency band can provide a better learning when it comes to the multistage learning process? Response to this question motives us to invoke a combination of the cluster-based and Fisher-based learning. In the combined method, a set of the frequency bands are firstly found in a way to provide optimal classification error. Among the sets of the frequency bands, exhibiting similar classification performance, the set with superior Fisher value is selected.

Referring to the previous sections, combining the two approaches of learning at the deep level can benefit the learning process of taking both the distribution and the classification rate into account. However, in the learning algorithm, the order of using Fisher value and the classification rate, as the constrained criteria, plays an important role in the learning process. In fact, the order of using the criteria depends on the designer if they prefer classification rate-oriented or discriminant-oriented. Putting the Fisher value as the primary criterion to be checked, attributes focus to the data distribution rather than the classification rate. Choosing the order of the criteria depends on the size of the training data and also the learning goal. This can be heuristically decided or empirically found through experiments.

8

Deep Time Growing Neural Network

Deep learning was first introduced in the architecture of the multi-layer neural network with a high number of layers and neurons, in a way to improve the learning capacity as was defined in the previous chapters. Time growing neural network was also introduced as an alternative to time-delayed neural network in which dynamic contents of an input time series is incorporated into a set of the temporal window with growing length, each characterized by its frequency contents, in conjunction with a multi-layer perceptron neural network for a nonlinear mapping. Invoking the growing time windows in any form of the growing scheme, can boost the performance of the classification method, compared to the time-delayed and multi-layer neural network, that was explained in the previous chapters. A time growing neural network inconsistently assigns the learning weights to the windows. This interesting feature of time growing neural network tailors the classification method optimal such that the windows with shorter lengths receives a higher number of the learning weights, and hence contents of the shorter windows are better learned compared to the longer ones. Besides, all the contents of the multi-layer perceptron and time-delayed neural networks are included in the time growing neural network, when spectral energies are employed for the learning task. Meanwhile, making the trade-off between frequency and time resolution is easier for time growing windows. All of these interesting features are included in time growing neural network, and several studies reported outperformance of this architecture compared to the other two alternatives, in the time invariant architecture [48][53][55]. Nevertheless, this architecture was described merely in the shift-invariant manner and needs to be improved by using a suitable structure of deep learning. This chapter is dedicated to describe a structure of deep learning as a dynamic process for time series analysis. Contents of this chapter are based on the studies, that have been newly published by the related journals or conference proceedings. In all of those studies, time growing neural network was employed as a building block of the learning process. The readers can ultimately speculate to suggest other structures of deep learning based on time growing neural network, which might be topics of the future publications.

8.1   Basic Architecture

This section unveils a general deep learning architecture for time series classification, based on the time growing neural network. This architecture assumes that a stochastic time series contains discriminative information for different classes, which are not easy to extract due to the non-stationary and non-ergodic behaviours, and meanwhile the discriminative contents are included within the signal chunks with short time, and also over a long interval of temporal variations. Hence, the architecture must be capable enough to learn both the short time information, and the long interval variation of the time series, named the dynamic contents of the time series. The deep learning architecture in this perspective, contains a block sophisticated for learning at the deep level, in conjunction with another learning process, which takes the long interval dynamics into the learning process. Although this architecture seems to be similar to multi-scale learning, but different in the sense of the dependence of the classification method as a whole to the learning methods at the deep level. Regardless of the dependence to deep level learning, multi-scale and deep architecture share common parts with each other in the learning perspective. Figure 8.1 illustrates a structure of the deep learning method for time series analysis with C middle classes. The number of the middle classes is indeed the state of the system and differs from the the number of the ultimate classes. The number of the middle classes must be often higher than the number of the ultimate classes, found either empirically or treated as a hyper parameter.

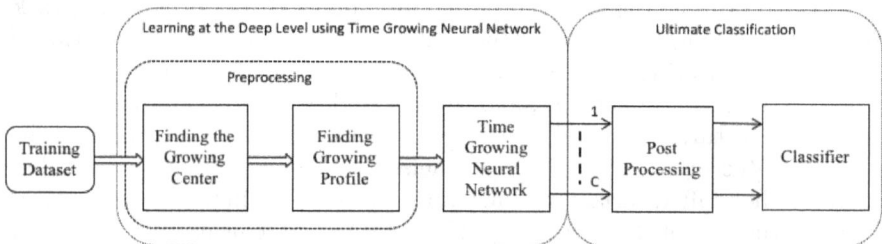

FIGURE 8.1: Overall architecture of a deep time growing neural network.

Before going through the method description, it is needed to draw the attention of the reader to the difference between temporal window and temporal frame, in the terminology of this book. Temporal frame is a piece of a time series identified either by following a systematic procedure or by performing a processing method, which might have an unfixed length, even though in many applications (including the below method) the frames accept uniform length. However, there is no obligation for a temporal frame to be attributed by a fixed length. An example where the temporal frames have variant length, is in speech processing, where the temporal frames encapsulate word utterance. This is an essential part of a speech processor, performed prior to learning semantics of the sentences. Depending on the word length, the corresponding temporal frame has a different temporal length. In contrast, the term temporal window always refer to a segmentation of time series in a well-defined manner. The temporal frames can have a variant length, however, the temporal windows cannot have unknown length. This is in line with the definition of temporal window in signal processing books. In this method, the time growing modules are invoked to learn time series contents within the short length temporal frames, referred to as learning at the deep level, which is usually followed by another level of learning that deals with dynamic contents of the time series. In fact, architecture of the input layer of the method is shaped according to the training data, keeping the rest of the architecture unchanged with respect to the data. By this definition, the learning process involves two different levels of learning, deep level learning and surface level learning. Deep learning corresponds to the process, through which detailed contents of the time series within short length temporal frames are learned over different classes of the training data. The method relies on the hypothesis that contents of the time series over a set of the short length temporal frames are presumed to be discriminative. Assuming that a temporal frame $x_{i,t}(n)$, of length T, is characterised by its frequency contents, calculated over a set of the temporal windows with growing length, all are situated within the temporal frame:

$$X_{i,t} = \frac{1}{T} \left| \sum_{t=0}^{T-1} x_{i,t}(n)e^{-j2\pi fn} \right| \tag{8.1}$$

The deep level of the learning, yields in a schematic of the growing windows, providing an optimal discrimination power. Besides, the frequency bands, over which the spectral energies, are found during the deep level learning. It is obvious that the frequency bands are accompanied by the temporal window, which is characterised by its spectral contents. An important question is the fashion of the growing windows to achieve optimal discrimination power, as was previously introduced in Chapter 4. It is sometimes the case that the discriminative contents of the short length frames are mostly concentrated within the beginning of the frame, and the last parts of the frame are rather consistent over the classes. This situation motivates the use of the backward time growing neural network, as the discriminative parts of the frame are employed with respect to the consistent part of the frame of the time growing scheme. In the deep time growing neural network, the type of the growing scheme (forward, backward, or bilateral) is learned through the deep learning process. The method, uses all the three forms of the growing windows, and finds the optimal windows in conjunction with the corresponding discriminative frequency bands. For the forward and backward cases, the scheme is rather clear, but for the mid growing scheme, the point from which the growing windows are bilaterally expanded, must be found as well. This point is named, the growing centre. The deep level of learning includes all these details. The following sequels describe the learning process.

8.2 Learning at the Deep Level

Although there are several interesting features of using the growing-time windows effectively which improve the learning process, when spectral energies are employed for the learning method to preserve the temporal variations, one should pay attention to the suitable scheme for using the growing-time windows in terms of the temporal expansion. Forward, backward and mid-growing schemes, each can improve the learning in its way. The question is, which scheme results in a better learning? In fact, the way of using the growing-time windows puts a noticeable impact on the learning performance, but it might be the case of degrading the performance, if implemented awkwardly. For the forward and backward cases, the length of the first time window within a frame certain frame length, plays an important role in the learning process, and so does the initial location and length of the fist window for the mid-growing scheme. Length of the first window, referred to as the initial window length, is treated as the design parameter for all of the three schemes, however, location of the initial window within a temporal frame is regarded as a learning parameter, often obtained in a non-iterative way. Learning at the deep level includes the process for finding the growing scheme as well as the discriminative frequency bands and the multilayer neural network, which will be described in detail, in the followings. Let's assume there are N time series with different length, in the training data, each has been divided into L_i temporal frame of length T_i. The time series contents at window t form the temporal frame i for a forward growing scheme can be formulated as follows:

$$\vec{x}_i(n,t) = x_i(n) : t = 1, ..., L_i \quad , \quad n = 0, ..., \left(t\frac{T_i}{L_i} - 1\right) \tag{8.2}$$

For the backward growing scheme the time series contents becomes:

$$\vec{x}_i(n,t) = x_i(n)\left(n + T_i - t\frac{T_i}{L_i}\right) : t = 1, ..., L_i \quad , \quad n = 0, ..., \left(t\frac{T_i}{L_i} - 1\right) \tag{8.3}$$

In the bilateral growing scheme, an initial temporal window is located at a point, named growing centre G, and grows from the both sides with a certain rate, until it can cover the signal. The growing centre can be identified by a percentage of the signal length, denoted by η in the following derivation:

$$\vec{x}_i(n,t) = x_i\left(n + \left(1 - \frac{t}{L_i}\right)G_i\right) : t = 1, ..., L_i \quad , \quad n = 0, ..., \left(t\frac{T_i}{L_i} - 1\right)$$

$$G_i = \frac{\eta}{100}T_i$$

$$\tag{8.4}$$

In the bilateral growing scheme, in addition to the initial window length, which is regarded as the design parameter, the middle point of the growing centre, is unknown and considered as a learning parameter (not a hyperparameter), found during the training process. The learning process involves a pursuit towards finding the discriminative temporal windows and frequency bands, in which the growing centre is obtained as well. The following sequels proposes an effective method for finding the growing centre. This method is, not the only one, but a strong alternative from a bunch of the heuristic ones, which may have been brought up in the readers' mind. The rest of the other learning elements will be described accordingly.

8.2.1   Learning the growing centre

If a temporal frame is divided into L_i segments, it is assumed that the growing centre is located in a temporal segment in which most of the information is included, compared to the rest of the other segments. One way to find this segment is based on using wavelet transform calculated for each of the temporal frames of the time series. Before explaining the learning process, wavelet transformation is briefly addressed. Wavelet transformation in a brief view, is a technique that provides a mathematical tool to decompose a time series into its constitutive components, where each of the components incorporates a certain range of frequencies [96][27]. The frequency ranges are fixed and ruled by the band-pass filters with the predefined shapes introduced by the wavelet type. Mother wavelet is a general term attributed with the wavelet transformation, exhibiting shape of the band-pass filters used for the decomposition. The frequency range of each component is a ratio of the sampling frequency, and the ratio is an integer factor of $\frac{1}{2}$ for dyadic wavelet transformation. The temporal frame can be decomposed into constitutive components by using discrete wavelet transformation technique:

$$X_i(\alpha, \tau) = \frac{1}{\sqrt{2^a}} \sum_{n=0}^{L_i} x_i(n)\psi_{\alpha,\tau}(n), \qquad (\alpha, \tau) \in Z$$

$$\psi_{\alpha,\tau}(n) = \frac{1}{\sqrt{2^\alpha}} \psi\left(\frac{n - \tau 2^\alpha}{2^\alpha}\right)$$

(8.5)

where $X_i(\alpha, \tau)$ is defined as the the wavelet transform of $x_i(n)$. The $\psi_{\alpha,\tau}(n)$ introduces the mother wavelet with the scale of τ and shift of τ to the transformation. Each value of τ is named a scale of the wavelet transform, carrying information over a certain frequency range. At each scale, the $X_i(\alpha, \tau)$ resulted from the derivation 8.5, is called the detail coefficients and if subtracted from the signal, the resulting function is the approximation coefficient. The above derivation shows that the sample rate is decreased by half of its value, for each increment of scale α. In order to obtain contents of the signal at each scale, the approximation and detail contents of the previous scale, are both up-sampled, meaning that a sample with zero value is added between each two samples. Then, the up-sampled signals are passed through a pair of filters, one for the detail and one for the approximation separately. The filters are designed according to the wavelet type, as defined by the wavelet family. The filters are named quadrature filters. This structure can be used to find the detail (high frequency contents) and the approximation (low frequency contents) of each scale. This process is named reconstruction of the signal. The reconstructed signals resulting from the detail and approximation of a scale, are then added together to create the signal at the higher scale. In many applications, the reconstruction is performed using detail contents only, keeping zero for the approximation. This yield of the details of the signal contents at each scale to be constructed as acts of filtering since parts of the frequency contents of the signal corresponded to the approximation contents are excluded from reconstruction. This technique has been widely used and explored by the researchers and scientists and dealing with more details of the technique is well-beyond the scope of the book, but it can be extensively found in a broad range of the literature [96][32].

Back to the main problem, finding the growing centre, the wavelet transform of the signal is employed for finding the growing centre. Assuming that the number of the temporal windows and the length of the signal is L_i and T_i, respectively. These two parameters are considered as the design parameters. The wavelet transform at each scale, is reconstructed up to the zero level, and denoted by $\widehat{X}_i(\alpha, \tau)$, where α is the scale number. The reconstructed contents of each time series, are divided into L_i non-overlapping temporal frames with fixed length of $\frac{T_i}{L_i}$. Energy of the wavelet transform is calculated and averaged for each temporal window, and also for each scale. The Fisher value of the energies are calculated for all the windows and scales, and the temporal window whose energy gives the highest Fisher value is selected as the growing centre. It is important to note that sometimes the energies must be verified first, before the Fisher calculations, in order not be lower than the noise limit. In these situations, one need to set a noise limit below which the windows are not invoked for the rest of the learning process.

8.2.2   Learning the deep elements

The deep elements of a time growing-based method are composed of the architecture to extract dynamic information from the time series, here at this point at the deep level. This means the deep contents extracted from the temporal frames. Figure 8.2 depicts a learning architecture for the deep level:

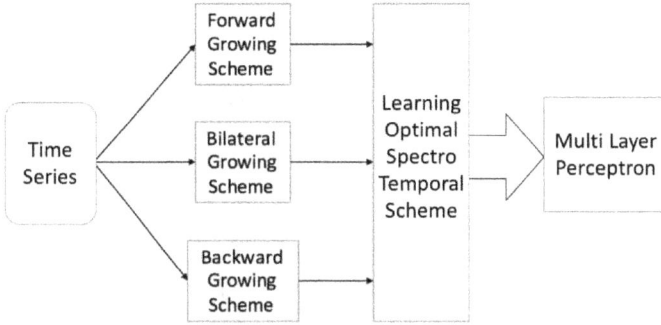

FIGURE 8.2: Learning architecture at the deep level for a deep time growing neural network.

This architecture contains the growing windows scheme as well as the discriminative frequency band for extracting spectral energies. The growing time scheme which was found during the training at the deep level, is found along with the discriminative frequency band, simultaneously, and the rest of the learning process corresponds to finding the learning weights to perform the nonlinear mapping, from the spectral energies to the classification result. We firstly deal with finding the discriminative frequency bands connected to the growing schemes, for calculating the spectral energies. To this end, we begin with spectral calculations. The spectral contents of each window are calculated using periodogram:

$$\vec{X}_i(f, t) = \frac{1}{T_i} \left| \sum_{n=0}^{T_i - 1} \vec{x}_i(n, t) e^{-2j\pi f n} \right| \tag{8.6}$$

The pursuit towards forming the input architecture of the classifier begins with considering the spectral contents and use of the three schemes of the growing windows; forward, backward and bilateral growing time window, for finding the discriminative time and frequency contents. The clustering method, introduced in Chapter 7, can be employed for the learning at the deep level. This level of the learning process leads to shaping the input structure of the deep learning method. This can be considered as the deep level learning. An important difference between deep time growing neural network and the other aforementioned deep learning methods, is that in deep time growing neural network, the pursuit is performed over the frequency bands in conjunction with the temporal windows and the joint frequency bands and the structure of the temporal windows are simultaneously found within the learning process. To this end, spectral contents of all the three schemes of time growing neural network together with the growing centre are found. The spectral contents are calculated for each temporal frame and averaged over the frames:

$$X_i(f,k) = \frac{1}{L_i T_{l,i}} \sum_{l=1}^{L_i} \left| \sum_{n=0}^{k\frac{T_{l,i}}{K}-1} \vec{x_{l,i}}(n,k)e^{-j2\pi fn} \right|^2 \tag{8.7}$$

$$X_i(f,k) = \frac{1}{L_i T_{l,i}} \sum_{l=1}^{L_i} \left| \sum_{n=0}^{n+T_{l,i}-k\frac{T_{l,i}}{K}} \vec{x_{l,i}}(n,k)e^{-j2\pi fn} \right|^2 \tag{8.8}$$

$$X_i(f,k) = \frac{1}{L_i T_{l,i}} \sum_{l=1}^{L_i} \left| \sum_{n=0}^{k\frac{T_{l,i}}{K}-1} \vec{x_{l,i}}(n,k)e^{-j2\pi fn} \right|^2 \tag{8.9}$$

In the above derivations, the learning process is followed by a pursuit over frequency bands f, temporal window number k, and the windows scheme s, where the later is an indicative number denoting the scheme of the growing window.

$$B \doteq \{(f_1, f_2, k, s) : (f_1, f_2) \in Z^+ \wedge f_1 < f_2 < F_{max}, k \in \{1, ..., K\}, s \in \{1, 2, 3\}\} \tag{8.10}$$

where

$$s = \begin{cases} 1, & \textit{Forward Growing Eq. (8.7)} \\ 2, & \textit{Backward Growing Eq. (8.8)} \\ 3, & \textit{Bilateral Growing Eq. (8.9)} \end{cases}$$

The spectral features are calculated over frequency band and the temporal windows identified by the windows number k and the growing scheme s. The growing scheme s points to one of the Eq. (8.7) to (8.9) to be used for the spectral calculation. For the bilateral growing case, the growing centre is supposed to be found as was described in Section 8.2.1, prior to the spectral calculation.

$$P_i = \left[\sum_{b_1 \in B} X_i(f,k), ..., \sum_{b_M \in B} X_i(f,k) \right] \tag{8.11}$$

The pursuit for finding the input architecture of a time growing neural network, employs a combined learning method at the deep level composed of the K-Means clustering and Fisher criterion. This way of calculation takes both the data distribution and discrimination power into account for the learning process.

Algorithm 10 The Deep Leaning Algorithm

1: **procedure** DLA($\langle x_{l,i}, c_i \rangle, K, \varsigma$)	
2: $b_1 \leftarrow \arg\max_{(b) \in B} \Delta(b) : S_W > TH$	▷ Eq. 7.13
3: **for** $i \leftarrow 2, M$ **do**	
4: **if** $S_W > TH$ **then**	
5: $B_1 \leftarrow \arg\max_{(b) \in B} \Delta^i$	▷ Eq. 7.13
6: $b_i \leftarrow \arg\max_{(b) \in B_1} I^i.$	▷ Eq. 7.2
7: **end if**	
8: **end for**	
9: **return** $\langle b_M \rangle$	
10: **end procedure**	

---

where TH is a certain threshold, set to secure the non-singularity of Eq. (7.13). This threshold is tentatively or sometimes empirically obtained by the developers.

The outcomes of this level of the learning are indeed a set of the frequency bands corresponding to their temporal windows with the growing schemes, whose spectral energies deliver an optimal segregation. This yields the first layer architecture of the classifier. As a matter of fact, these outcomes constitute the input node of a multi-layer perceptron neural network, trained by using back propagation error method, as was previously introduced. It is typically sufficient for the neural network to contain three layers of neurons along with the input node which is built up for learning time series contents. There is no restriction on the number of the layers as well as the number of the neurons. The activation functions for the hidden and output layers can be tangential sigmoid and logarithmic sigmoid, respectively. Nevertheless, objectivity of the method is not lost by using other activation functions. The number of the neurons in the hidden layer cannot place an impact on the learning process, after a certain number, however, if the number of the neurons decreased down to a certain threshold, the learning process may be affected [55][48]. The number of the neuron at the output layer reflects the level of deep learning, and can be intuitively found by considering possibility of the pattern occurrence in the time series.

8.3 Surface Learning

Outputs of learning at the deep level, shape up a multidimensional time series of real number. Therefore the results coming from this level of learning, at the deep level, will be processed to give the final classification result. The multidimensional time series resulting from the deep processing, conveys discriminative information about the time series contents at the deep level. In fact, this processing extracts the deep discriminative contents of time series from the deep level, and brings them to a surface for ultimate learning. This is the main motivation for naming the processing as the "surface learning". In many practical cases, a simple post processing in conjunction with a nonlinear classification such as multi layer perceptron classifier can sufficiently cope with the learning task at the surface level. Some others invoke another dynamic processing methods on these outcomes. The need of the dynamic learning methods such as recurrent neural network or dynamic time warping can be obviously felt in many applications, where the time series reveals further complicated features. Such dynamic methods can be easily incorporated into the learning model for the time series classification. This time the elaborated dynamic methods can takeover the role of the post processing and the ultimate classification, which will result in an accurate performance. Nonetheless, for simple cases, in the post processing, sometimes an unsupervised learning such as K-Means is invoked to exclude unimportant contents, i.e., noise contents, from the learning process. In any case, if the dynamic contents, can contribute in the surface level of learning, to improve the classification performance, a recurrent model might be invoked that suits the dynamic surface learning. In this case an intuition about the dynamic contents at the surface level is always needed. As like as other cases, for the dynamic cases where the sequences are well segregated in the space, but not easily seen in the multidimensional time series, dynamic time warping can be always employed. Nevertheless, this method demands a long processing time, even at the test level which makes the implementation, far more complicated than the recurrent neural networks when it comes to the practical application. A fast method for implementation of dynamic time warping has been on demand for such the processing. As you will see in continuation, the idea of recurrent neural network is based on quick learning, as with any other neural network-based methods, unlike the statistical methods such as hidden Markov model or dynamic time warping. This would be at the expense of lower structural risk for such the statistical classification methods. The readers are encouraged to review the statistical model-based classification methods in light of the learning and testing complexities as well as the structural risk, comparing to the neural network-based methods [44][43].

9

Deep Learning of Cyclic Time Series

Real-world activities in many cases are examples of cyclic activities, resembling cyclic time series, when recorded by appropriate instruments reflecting the behaviour of the activities. It is evident that a level of noise is always associated with the recording due to many factors, such as electrical induction and thermal condition. In biological activities such as heart sound signal, even more factors come into the phenomenon such as biochemical status of the blood, heart rhythm and respiration, all constituting different sources of the noise, that can resemble cyclic characteristics, even at the presence of the noises. One should pay attention to the difference between periodic and cyclic time series in terms of their natures and behaviors. Periodic time series, is an absolute and impractical definition with theoretical applicability, however, in an engineering point of view, some of the activities appear as having periodic contents with sufficient approximation. The activity of a clock pulse generator in microprocessor systems is an example of being treated as a periodic signal. Although the period of the generated signals takes an approximation in the range of 0.000001 hertz, in the engineering context this would be assumed as zero. It is therefore, a question of application. Nevertheless, the term "periodic signal" is an absolute term that has modest roots in reality. In contrast, cyclic time series is an adaptation of the periodic characteristics which can fit well into real-world applications. A cyclic time series is not periodic, but instead certain patterns repetitively appear in certain intervals, known as the cycles. The cycle duration is not a fixed number, in contrast with the absolute definition of periodic signals. Contents of a cyclic time series within the cycles resemble stochastic behavior, which associate even further complexities with the learning methods. In this chapter, practical methods are introduced for learning cyclic time series. The time series is assumed to be all end-pointed, meaning that the beginning and end points of the cycles are all identified by other methods. There are several methods which are sophisticated for the segmentation of cyclic time series, especially for the time series of heart sound signal [121][49], mainly working based on the energy difference of the time series over sets of the temporal windows. As such the methods have been well incorporated into different learning methods such as automatic methods for speech recognition, and natural language processing. Appropriate machine learning methods for learning this group of the time series are urged to consider learning non-stationary stochastic contents within the cycles, along with the dynamic variations over the cycles. This may remind the readers the topics of "learning at the deep level" and the "surface learning", from Chapter 8. Looking back and remembering from Chapter 8 the learning method at the deep level, employs temporal frames of fixed length to extract the deep information. Outcomes of this learning level are later employed by a dynamic classification method to perform the surface learning. Likewise for the cyclic time series, the learning method invokes two different levels of learning, when it comes with the cyclic stochastic time series: cyclic learning and surface learning. Figure 9.1 illustrates a block diagram of the learning process for the classification of cyclic time series:

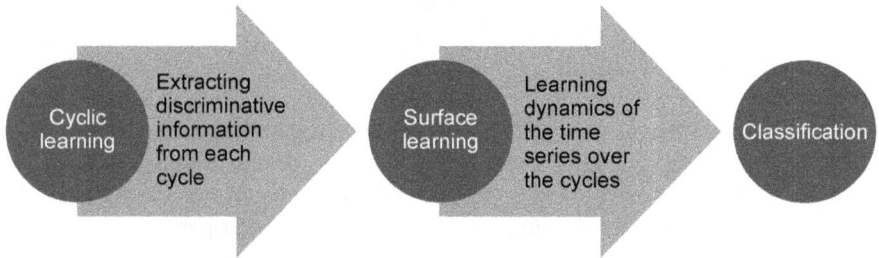

FIGURE 9.1: Classification of cyclic time series demands two levels of processing: learning within-the-cycle contents (cyclic learning) and learning over-the-cycle contents (surface learning).

It is shown in the above illustration that cyclic learning performs a kind of the learning at the deep level. However, there is a big difference between the cyclic stochastic learning and the non-cyclic one; as the cycles do not have a equal time period, deep learning cannot be designed based on the temporal frames with a fixed length. In fact the temporal frames are replaced by the cycles, and the surface learning remains typically identical to the previously described method. In order not to lose objectivity of the description, the cyclic learning employs sector analysis instead of windowing the temporal frames, as was the case for the non-cyclic stochastic time series. The following sections will take this point into account and introduce different methods for processing cyclic stochastic time series in light of the learning for the classification problems.

9.1   Time Growing Neural Network

In the previous chapters, we sow that learning process at the deep level for cyclic time series is slightly different from the non-cyclic time series. A non-cyclic time series lacks from any repetitive pattern and therefore, a uniform temporal frame can be employed for the learning process. This gap between cyclic and non-cyclic time series can be covered by taking the sector definition into account, as was addressed in Chapter 4. This brings a level of deep learning, named cyclic learning to the process. The cyclic learning in the training phase involves two steps of deep learning: finding the input architecture of the time growing neural network and finding the learning weight of the neural network. It is assumed that all the cycles were already end-pointed, and $T_{l,i}$ denoted length of the cycle l in time series i. Using training data of size N, at the first step, a time growing scheme is shaped up, using the temporal sectors instead of temporal frames. One of the design parameters at this level is the number of the sectors in each cycle, denoted by K, which is in turn an indication of the temporal resolution. Higher resolution is used, the better resolution it yields, at the expense of the increased complexity and longer training time. A reasonable range of K is often intuitively obtained by considering the temporal and spectral characteristics of the learning data, and the optimal number is found through the optimisation process. Spectral contents of each sector, for the forward, backward, and bilateral growing schemes can be obtained by the Eqs. 9.1, 9.2, and 9.3, respectively, using periodogram:

$$X_i(f,k) = \frac{1}{L_i T_{l,i}} \sum_{l=1}^{L_i} \left| \sum_{n=0}^{k\frac{T_{l,i}}{K}-1} \vec{x_{l,i}}(n,k)e^{-j2\pi f n} \right|^2 \tag{9.1}$$

$$X_i(f,k) = \frac{1}{L_i T_{l,i}} \sum_{l=1}^{L_i} \left| \sum_{n=0}^{n+T_{l,i}-k\frac{T_{l,i}}{K}} \vec{x_{l,i}}(n,k)e^{-j2\pi f n} \right|^2 \tag{9.2}$$

$$X_i(f,k) = \frac{1}{L_i T_{l,i}} \sum_{l=1}^{L_i} \left| \sum_{n=0}^{k\frac{T_{l,i}}{K}-1} \vec{x_{l,i}}(n,k)e^{-j2\pi f n} \right|^2 \tag{9.3}$$

where L_i is the total number of the cycles in time series i. Using the $T_{l,i}$ as the cycle length, $x_{l,i}$ is defined for the the forward, backward, and bilateral growing schemes by the equation (8.2), (8.3), and (8.4), respectively. Use of the growing **sectors** scheme, introduces an inconsistency to the length of the temporal windows, which can in turn introduce inconsistent frequency resolution. It is obvious that one way to cope with this problem is zero-padding to a predefined length, that guarantees all the windows attain identical length after the zero-padding. Even though there is no theoretical restriction to use the sectors with a different length, when mapping a cycle to the spectral domain, the zero-padding is needed when it comes to averaging over the cycles in order to maintain concordance of the frequency samples. Here, it is worth noting that use of the growing sectors provides further flexibility to preserve temporal dynamics in the learning process rather than wavelet transformation, as the temporal shrinkage in wavelet transformation is performed by a certain multiplication of the scale number. The learning process at the deep level is similar to the non-cyclic learning, with this difference that the spectral contents in (8.6) is replaced by the above equations. The learning algorithm as well as finding the discriminative frequency bands, sector and temporal schemes remain similar to the previous sections.

9.2 Growing-Time Support Vector Machine

The main idea of incorporating different schemes of the growing time window into a classification architecture is to build an input layer which enables the architecture to learn discriminative features at the deep level of the learning process. The surface learning would be performed by a binary classifier such as a multilayer neural network. The binary classification can be performed by support vector machine. In this case, the spectral energies, extracted from the temporal sectors, constitute feature vectors for each cycle, carrying information from the deep level of learning within the cycle. Figure 9.2 illustrates a block diagram of a deep learning method for cyclic stochastic time series for a problem of binary classification. In this method, firstly a deep learning process is employed to learn each class versus the others using support vector machine. This process is known as "singling-out" learning. In this process subtle contents of the "within-cycle" are learnt for each cycle, which requires multiple learning, individually for each class of data. This multiple learning requires data preparation for each individual learning, C times (C is the total number of the classes), where a label 1 is assigned to each sample of the learning data from that individual class, and 0 for the rest. This is repeated C times with each sample receives the labels 1 only once. Next, discriminative frequency bands and the rest of the the deep learning process is performed using a support vector machine [47]. Lastly, surface learning is performed by using any classification method. In a binary case, support vector machine can be invoked. The learning manner is detailed as follows:

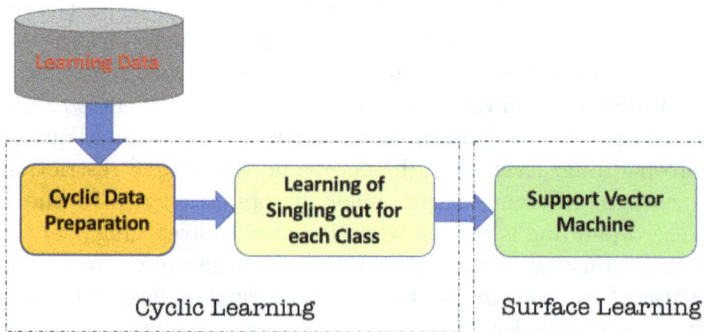

FIGURE 9.2: Deep learning for cyclic time series using support vector machine.

This process is explained for the training phase first, followed by the testing phase consequently, in a separate paragraph. Considering a case of binary classification, a hyperplane $\Phi : R^K \mapsto \{0, 1\}$ that can separate the two classes is found by using support vector machine:

$$y_{l,i} = \Phi(P_{l,i}(X_i(f, k)))$$
(9.4)

where l is the cycle number ($l = 1, ..., L_i$), and i is the subject number ($i = 1, ..., N$). L_i is the total number of the cycles exist in the subject i. Details of finding the hyperplane are found in [23][9][141]. In these conditions, surface learning is not simply limited to a one step classifier, such as neural network. One way to achieve a robust classification is based on using a level of post processing cascaded by the binary classification. An important assumption to this method is that the surface learning is not attempted to extract dynamic variations over the cycles. Instead, it treats the over-cycle variation as caused by noise contamination, where statistical manipulation can remove the noisy outcomes. In order to eliminate the effect of the noise from the outcomes of the cyclic classification, average of the cyclic classification is calculated by the following formula:

$$O_i = \sum_{l=1}^{L_i} \Phi(P_{L_i}(X_i(f, k)))$$
(9.5)

where O_i is the average of the cyclic classification. In the training phase, membership of $y_{l,i}$ to the both of the classes are found independently, by using statistical estimation. Gaussian distribution is a typical membership function used in many cases of statistical processing, especially when the size of the training data is sufficiently large. There are iterative methods for finding the statistical distribution of the real data, however, in this case since the result of the classification can take over parts of the learning process, a rough estimation would be sufficient in the majority of the cases. This approximation can be based on the average and variance of the classification outcomes at the deep level, or alternatively the cyclic learning is employed for approximating the membership functions. The membership values of the signals to different classes are then invoked for surface learning, which yields the ultimate classification result. Depending on the case study, any kind of classification method can be used for surface learning. A rule of thumb valid here says that for binary classification problems with small or medium data size, support vector machine is always a recommended option, and for the multi-class, multi-layer perceptron neural network is a common candidate. All the above-mentioned sequences correspond to the training phase. The testing phase is rather straightforward, and demands only simple calculations, performed on the obtained parametric functions from the training phase.

9.3 Distance-Based Learning

The previous cyclic learning methods were based on a phase of learning at the deep level, herein named cyclic learning, through which a set of the discriminative frequency bands along with the sectoral schemes were obtained. The presented methods so far, invoked spectral contents of the temporal sectors with growing length for the spectral estimation. Such is the fashion of the temporal windows to provide an efficient way of learning by prioritizing the temporal sectors according to their importance. The temporal sectors with the highest distinction receives the highest number of the learning weight, in another term, its contents are learned the best. However, this is only valid when certain structures of the learning parameters such as neural network or support vector machine are employed for the training, which is not the case for the methods which are based on distance measurement, like KNN. Regardless of the sector fashion used for training at the deep level, spectral contents of the sectors are invoked for learning at the deep level. The spectral contents calculated over the sectors, in any kind of fashion of growing or non-growing fashion, are named spectro-sectoral contents of the time series. Using these spectro-sectoral contents, the discriminative frequency bands, are found using any algorithm such as the hill-climbing method, as was described before in different contexts. These bands all together, help the method to profile dynamic contents of the time series within the cycles. In the training phase, the spectro-sectoral contents of the time series are obtained by using periodogram, calculated over a set of the temporal sectors with fixed number of the sectors per cycle. The discriminative frequency bands can be obtained using Algorithm 10. Then, a KNN method is employed, through which the distance of the KNN to each class is obtained. As with the previous methods of deep cyclic learning, membership functions are calculated over the result of the cyclic classification, performed using a KNN method. The membership values are employed by any kind of the classification method, such as neural network, which provides a nonlinear mapping, and the learning weights are found by any kind of the methods. The testing phase involves easy calculation of the spectro-sectoral contents over the discriminative frequency bands, and the functions obtained in the training phase.

9.4 Optimization

Learning processes of any kind, incorporates a phase of training followed by optimization, where the former delivers the learning parameters and the later gives the design parameters. The number of the temporal sectors as well as the length of the temporal windows for the spectral calculation are the two design parameters associated with the above-described methods. The design parameters cannot be uniquely obtained with such methods. It is customary to find a range for the parameters according to the nature of the time series. For example, range of the sampling frequency as well as the sector length are roughly approximated for the distance/based method, and then a statistical procedure is invoked for finding the optimal set of the design parameter. At this point, there is a big difference between large and small data cases of learning. For the large data cases the attempts are focused on including various samples of the learning data, while for the small cases running around from overfitting is important issue. A procedural algorithm is suggested by calculating performance of the methods using different set of the design parameters. The A-Test method can be invoked either with or without premutation (see Chapter 2.10.3). It is obvious that permutation imposes complexities to the method, and increases the execution time and the memory issues. A smart selection can be based on the verification of the effect of the permutation on the performance of the method. Any reader can arrive at a conclusion to use a heuristic or sometimes adhoc method suiting the case study.

10

Hybrid Method for Cyclic Time Series

The context of time series analysis has received a large number of the model-based learning methods for the purpose of classification in the last three decades. In contrast with knowledge-based methods in which information is extracted directly from the data, model-based methods presume a structure of the data flow in conjunction with a set of the hypothesis and parameters, whose values are found by the learning method. In fact, the outcome of a learning method is a set of values for the learning parameters, providing an optimal performance for the learning task, which is tightly dependent on training data. Hidden Markov model and artificial neural networks are known as the two model-based learning methods, whereas methods like case-based reasoning and fuzzy logic in which information is extracted directly from the data without the need of presuming a specific model generating the time series. Knowledge-based learning is far from the main focus of this book, even though fuzzy method was addressed to be as an especial tool for deep learning. The two model-based methods, which were earlier brought up into the discussion, hidden Markov model and neural networks, differ from the point of data generation. In hidden Markov model, the time series contents are generated according to the probabilities, which attribute a self-generative feature to the model, however, neural networks cannot produce time series by itself, and the output depends only on the inputs fed to the model. This comes from the fact that the model is hidden for the former, implying that a time series can be generated by itself using that model, and an input time series is classified according to the probability of belonging to a hypothesised model that produces the time series. Neural networks, in general, employ posteriori data for training, in contrast with the hidden Markov model where the initial probabilities together with the priori probabilities play important roles in the learning process. Nevertheless, one should consider that even for neural networks, lack of data from certain classes can be roughly regarded as an indication of the priori probabilities. In any case, the initial symbol and state probabilities are not considered in the parameters of a neural network, causing the inability of the neural network for self-generating a time series in its classical way. One can certainly propose heuristic model based on neural network to generate time series. Nevertheless, one cannot overlook the very elaborating features of neural networks in quick learning with relatively small amount of data. It was therefore, rational that many studies in the 1980's, were directed towards combining the two methods, for improved learning, which were named hybrid models [149][82][126]. Application of the hybrid models have been highlighted in different field of study, especially in the field of speech recognition, in which enhancement in the recognition was practically observed [157][139][39][31]. Links between neural networks and hidden Markov model in terms of estimating posteriori probabilities was discovered in that decade. The model-based methods can effectively learn that grammar exists in time series data. This grammar comes from the semantic relation ruling the sequential data. In speech processing, a hybrid model, is assigned to a model which combines neural network and hidden Markov model, however, generality of this definition is not lost if other combinations are employed, even though this way of terminology is more common is this topic. In this chapter, a specific hybrid model of deep learning for cyclic time series is introduced. It is worth noting that another important difference between hidden Markov model and deep learning methods is that a level of pre-processing is performed in hidden Markov model-based methods to turn a time series of real valued numbers into another form of data: time series of patterns, and the recognition task of the Markov model is performed using the time series of patterns, while the deep learning methods usually do not involve a heavy level of the pre-processing and the whole process of the recognition is performed in an integrated manner inside the architecture of classifier. However, the following method employs a combination of the conventional hidden Markov model and neural networks, where a pre-processing level is also employed. Figure 10.1 shows block diagram of the method:

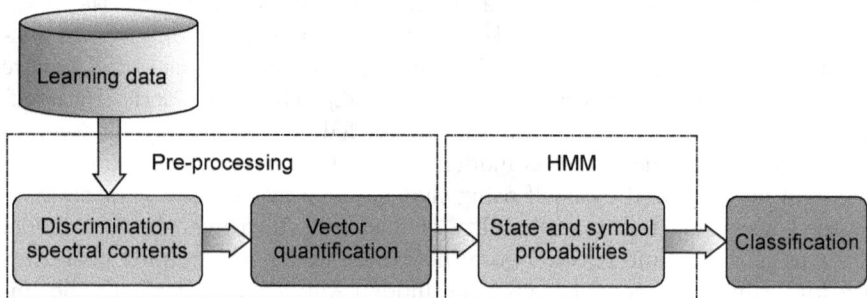

FIGURE 10.1: Classification of cyclic time series using hybrid model. A preprocessing level, composed of extraction of discriminative frequency bands and vector quantification is performed and followed by the Hidden Markov Model (HMM). State, state transition and symbol probability at each state are calculated and employed for the classification.

The preprocessing serves as the learning at the deep level, as was discussed in the previous chapters, however, with this difference that all the time series of the training data show cyclic behaviour and have therefore already been end-pointed and segmented. Therefore, the time series contents are segmented for each cycle, and labeled for each state. As with other hidden Markov model-based methods, the following assumptions are made for the learning process all throughout this chapter:

- N: The total number of the training data

- L_i: The total number of the cycles exist in a time series i

- $x_{l,i}$ ($l = 1, ..., L_i$ $i = 1, ..., N$): A time series for which l denotes cycle numbe

- $T_{l,i}$: Cyclic interval

The above definitions incorporate only the time series, or the inputs, and does not address the method or any design parameter. A Markov model is often expressed by its state transition model in conjunction with the probabilities. Since we employ spectral contents of time series for the learning process, the spectral energies are also considered as a part of the design method. Here, attention must be paid for the spectral calculations, when temporal windows of fixed length are employed, and the length of the temporal windows is regarded as a hyper parameter for the method. Based on this assumption, we have:

- $X_{l,i}(f, t)$: Spectral contents of the time series at frequency f and temporal window t, defined with respect to the beginning of cycle l.

- Λ: Length of the temporal window for calculating the spectral contents ($\forall(l, i)$: $\Lambda << T_{l,i}$).

- $s_{l,i}(t)$: State of the time series i at time t and cycle l.

- $\mathcal{A}$ and $\mathcal{B}$: Symbol and state probabilities.

The method description begins with an introduction to learning at deep level, followed by the rest of the classification process. This method is clearly introduced not only as an absolute model for learning, but to make the readers acquainted with the recent methods as an introduction for inspiration. After this chapter, the readers may become inspired enough to bring in innovative methods, according to the case studies of their underlying researches.

10.1 Learning Deep Contents

Learning contents of a time series at the deep level intrinsically demands different strategies for cyclic and non-cyclic time series. This level of learning corresponds to an autonomous procedure for feature extraction, which is slightly different from the procedures that were presented in the previous chapters in regard with the time growing-based methods. Sector analysis, is in fact the main point of diverse in the sense of processing methodology, that is considered in cyclic time series only, when it comes with the time growing fashion for the learning purpose. This process leads to extraction of the similarities and dissimilarities, termed by learning process, based on the spectral contents of time series which are extracted from the growing time windows. The number of the growing windows is treated as a design parameter, which in turn governs length of the temporal windows, and hence the learning capacity. In contrary, for the non-cyclic time series, temporal windows of fixed length can provide efficient means for learning dynamics of time series, as in hidden Markov model. This manner of learning the dynamic contents can be invoked even for cyclic time series, specially when the cycle durations are sufficiently long, and the time series is resulted from a model of the sequence with pre-defined states. An important presumption for the hybrid method is to assume a hypothetic model that creates the cyclic time series in which the state sequences were already identified for each time series in a direct or indirect way. Although each cycle can include several states, the spectral calculation is performed over the whole cycle and deviation of spectral energies within each state is considered as a key learning feature. With this perspective, the cyclic frequency contents are directly obtained using periodogram:

$$X_i(f) = \sum_{l=1}^{L_i} \frac{1}{T_l} \left| \sum_{n=0}^{T_i-1} x_{l,i}(n)w(n)e^{-2\pi fn} \right|^2 \tag{10.1}$$

where $W(n)$ is the temporal window employed for attenuation of side lobe effect. Zero padding is performed in order to achieve consistent frequency resolution for all the cycles. Assuming that a frequency band is defined as in Eq. (10.2), spectral feature can be calculated by Eq. (10.3):

$$B \doteq \{(f_1, f_2) : (f_1, f_2) \in Z^+ \wedge f_1 < f_2 < F_{max}\} \tag{10.2}$$

$$P_i = \left[\sum_{b_1 \in B} X_i(f), ..., \sum_{b_M \in B} X_i(f) \right] \tag{10.3}$$

where F_{max} is the maximum frequency presumed for the spectral calculation. F_{max} can take a maximum value of $\frac{F_s}{2}$ (half of the sampling frequency) which is known as the Shannon-Nyquist frequency [105][104][32]. The learning process at the deep level is indeed based on finding the discriminative frequency bands which can provide optimal discrimination Algorithm 11 is one way to obtain the discriminative frequency bands.

Output of the algorithm is constituted of a set of the frequency bands which provide optimal discrimination for each state. These bands are employed by the method to transform an input time series to another domain of spectral energies which is a time series of the multi dimensional feature vectors. In fact, this level of the learning process, applies a nonlinear mapping to the time series. The resulted time series is another time series composed of the spectral energies. The time series resulted from this learning process will be

---

Algorithm 11 The Learning Process at the Deep Level for Hybrid Model

---
1: **procedure** DLA($\langle x_{l,i}, c_i \rangle, K, \varsigma$)
2:      $b_1 \leftarrow \arg\max_{(b) \in B} \Delta(b) : S_W > TH$ ▷ Eq. 7.13
3:      **for** $i \leftarrow 2, M$ **do**
4:          **if** $S_W > TH$ **then**
5:              $B_1 \leftarrow \arg\max_{(b) \in B} \Delta^i$ ▷ Eq. 7.13
6:              $b_i \leftarrow \arg\max_{(b) \in B_1} I^i.$ ▷ Eq. 7.2
7:          **end if**
8:      **end for**
9:      **return** $\langle b_M \rangle$
10: **end procedure**

---

invoked by the second stage of the learning process, introduced as the process of cyclic learning.

10.2   Cyclic Learning

Cyclic learning for hybrid method involves the following mappings:

- $x_i(n) \rightarrow X_i(f, t)$: From a one dimensional time series of $R^{1 \times L_i T_i}$ to another time series of different length, but multi dimensional of $R^{F_{max} \times L_i \Gamma_i}$, composed of the spectral contents. Γ is the total number of the temporal windows used for the spectral calculations

- $X_i(f, t) \rightarrow P_i(t)$: From a multi dimensional time series of $R^{F_{max} \times L_i \Gamma_i}$ to another multi dimensional time series of $R^{M \times L_i \Gamma_i}$ composed of the spectral energies which are calculated over M discriminative frequency bands ($M << F_{max}$)

- $P_i(t) \rightarrow O_i(t)$: From a multidimensional time series of $R^{M \times L_i \Gamma_i}$ to a single dimensional of $Z^{1 \times \Gamma_i}$ composed of the the numerical symbols resulted from the vector quantification of the spectral energies

- $O_i(t) \rightarrow V_i$: From a single dimensional time series of $Z^{1 \times \Gamma_i}$ to a multidimensional feature vector of R^{Υ}, composed of the state and symbol probabilities

- $V_i \rightarrow q_i$: From multidimensional vector of R^{Υ} to a number denoting class of the time series

In the above mapping items, the first and the second items are performed by Eq. (10.1) and Eq. (10.3), respectively. It is obvious that the spectral energies in (10.3) are calculated over the discriminative frequency bands, obtained using Algorithm 11. The third item requires a vector quantification method such as Euclidean distance D_E (Eq. 10.4) or Mahalanobis D_M distance (Eq. 10.5):

$$D_E(P_i(t)) = \sqrt{(P_i(t) - \mu_c)^T(P_i(t) - \mu_c)} \quad , \quad c = 1, ..., C$$
$$O_i(t) = \arg_c \min D_E(P_i(t))$$

(10.4)

$$D_M = \sqrt{(P_i(t) - \mu_c)^T \Sigma_c^{-1}(P_i(t) - \mu_c)} \quad , \quad c = 1, ..., C$$
$$O_i(t) = \arg_c \min D_M(P_i(t))$$

(10.5)

where μ_c and Σ_c are the mean vector and covariance matrix of data for the class c. This quantification can be performed by using other methods such as KNN or even multi layer perceptron neural network. The probability feature vector in the forth item is then calculated by:

$$V_i = [\langle E(s_{l,i}(t))\rangle, \langle E(s_{l,i}(t2)|s_{l,i}(t1)))\rangle, \langle E(c|s_{l,i}(t)))\rangle]$$

(10.6)

The feature vector found in Eq. (10.6) will be employed by a classifier to perform the ultimate classification as shown in the fifth item. This part demands more sophistication which will be described in the following section.

10.3   Classification

The previous mappings ended with a multidimensional feature vector, associated with an input time series which preserves dynamic contents of the time series, reflected by the probability features. The probability features include symbol, state, and state transition probabilities, all together conveying dynamic of the input time series. The feature vectors are supposed to be discriminative for all the classes. The ultimate classification can be performed by any static classifier, such as multilayer neural network, support vector machine or even distance-based methods such as KNN. As a rule of thumb, for the problems of binary classifications with small and medium size of the training data, support vector machine is an excellent option, where an appropriate kernel function can be empirically selected based on the intuitive insights into the data distribution. In many applications linear kernel function works well, however, for the cases when the classes are tightly close to each other, nonlinear kernel functions like quadratic or Gaussian function can be considered as the relevant options. For the cases with large data, a multilayer neural network is recommended, with a training method which includes batch training. A favourable feature of neural networks is the flexibility of the learning process with a different size of training data, as well as its well-known feature: quick learning. Other classification methods such as Fuzzy classifiers and KNN method are typical alternatives which must be investigated for any classification problem. There is no specific rule for selecting an appropriate classifier and experience of the developer plays an important role for selecting an appropriate classification method, in the validation process, which is an important part of it's development.

11

Recurrent Neural Networks (RNN)

11.1   Introduction

The neural networks discussed in the preceding chapters, are purely static learning models that implement mapping from the input vector to the outputs. However, they cannot deal with time related information in a dynamic system. Consider the industrial process of gas furnace as an example, the CO_2 concentration is not only dependent on the current air flow rate but also affected by the process conditions in the preceding steps. Another example can be seen from data analysis with EEG signal in which the temporal property of the data plays a key role in the pattern recognition of the signal. Evidently, handling the dynamic nature of data sequences requires deep networks other than those traditional static learning models.

Inspired by the self-feeding mechanisms inside human brains, recurrent neural networks have been developed, which resort to a recurrent architecture to capture the time-dependent associations. They can be used to process sequential data and predict future data values by incorporation of the influences of previous data evolutions [99]. Recurrent networks are shown as universal approximators to dynamic systems [1][2] and also scalable in learning how much longer sequences than those manageable by the feedforward networks without sequence-based treatments.

There are generally two types of recurrent networks: globally recurrent and locally recurrent [3]. Networks of the former case allow for feedback connections for any pairs of neurons, while locally recurrent networks contain dynamic inner feedback despite the globally feedforward connections between neurons. A fully connected recurrent network with neurons of sigmoid activation functions has equivalent computational power to that of any state-space machines [4]. However, the learning of globally recurrent networks may be complicated by the problem of instability. Locally recurrent networks are relatively simpler for training due to less complex architecture. Yet they are able to approximate the state-space trajectory of a continuous function with any desired accuracy [5].

The early works of recurrent networks dated back to the models proposed by Elman [6] and Williams and Zipser [7] independently, which marked the start of the research of applying neural network techniques in processing temporal information. Later various recurrent networks with different structure and functioning were proposed, although the notion of "recurrent neural network" has not been clearly defined in the literature. More recently, in relation to the avenue of deep learning, (deep) recurrent networks are being investigated to learn the long-term dependence from longitudinal time series signals.

This chapter focuses on local recurrent networks as models in the state space, in which hidden neurons correspond to states of an underlying process and the value of a hidden neuron is affected by the states in the preceding time step. The learning of these networks is based on the fundamental principles of structure unfolding in time and parameter sharing in the network. Timely unfolding enables converting the temporal model into a multilayer feed-forward network so that existing static learning techniques can be reused. The idea of parameter sharing manifests in applying the same model for inference at all time steps. Each member of the network output is produced using the same function as previous outputs. Likewise, the values of hidden neurons are updated with the same rule at different locations of the sequence. Therefore the unfolded network consists of identical sub-models that are connected through consecutive time steps.

The rest of this chapter proceeds as follows:

Section 11.2 describes (local) recurrent neural network as a state space model, which is followed by unfolding the network with time in Section 11.3. Section 11.4 explains the backpropagation through time (BATT) algorithm for the learning of recurrent networks. The challenge of learning long-term dependencies is discussed in Section 11.5. In Section 11.6, we present the long short-term memory (LSTM) model as an extended recurrent network to overcome vanishing gradients. Finally, Section 11.7 provides an outline of other types of recurrent networks.

11.2   Structure of Recurrent Neural Networks

Let's consider a dynamic process as depicted in Figure 11.1, in which X and Y denote the external input and output vectors, S represents the vector of hidden states, and the square indicates the delay of a time step. The state transition and observation functions of this process are formulated in Eqs. (11.1) and (11.2) respectively.

$$S_t = F(S_{t-1}, X_t) \tag{11.1}$$

$$Y_t = G(S_t) \tag{11.2}$$

The state transition from S_{t-1} to S_t is seen as autoregressive as it is also affected by the external force X_t. The output Y_t is calculated in the observation function based on the current state S_t.

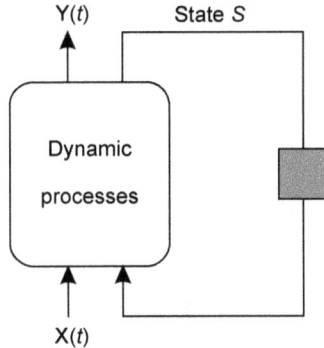

FIGURE 11.1: A dynamic process with state S.

Recurrent neural networks, as universal approximators, can be well utilized to represent temporal dependences of dynamic processes. More specifically, this chapter focuses on local recurrent network with one hidden layer as the dynamic process model. The architecture of this network is shown in Figure 11.2, where the recurrence only exists with the hidden neurons $s_1, s_2, \ldots, s_d$, which are fed back to themselves in the next time step. This means that the hidden units are determined with non-linear functions of their previous values and the external forces, while the network outputs are calculated only based on the current values of the hidden units. Thereby, the state transition and process outputs can be modelled by the equations as follows:

$$S(t) = tanh[W \cdot S(t-1) + U \cdot X(t) + \theta] \tag{11.3}$$

$$Y(t) = V \cdot S(t) + \gamma \tag{11.4}$$

where $S = (s_1, s_2, \ldots, s_d)$, U, V, and W are the weight matrices of the connections of the network, and q and g are the bias vectors for the hidden and output units respectively. Note that, for classification problems, a softmax operation will be applied as a post-processing step to $Y(t)$ to obtain the normalized probabilities of possible classes.

The recurrent network in Figure 11.2 is a succinct graph. It actually implements a function that takes the sequence of the external forces X till time step t to produce the network outputs $Y(t)$. This is evident by repeated applications of the mapping $h : X(t) \times S(t-1) \to S(t)$ across various time steps as follows:

$$\begin{aligned} Y(t) &= h[X(t), S(t-1)] \\ &= h[X(t), h[X(t-1), S(t-2)]] \\ &= H[X(t), X(t-1), \ldots, X(1)] \end{aligned} \tag{11.5}$$

Training the network is equivalent to adapting the transition function F in (11.1) and observation function G in (11.2) of the dynamic process. It can

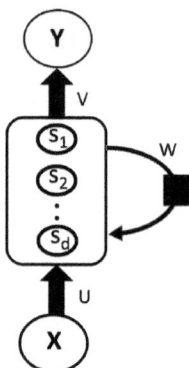

FIGURE 11.2: A recurrent neural network with d hidden neurons.

be transformed to the following parameter optimization problem concerning the weight matrices U, V, W and bias vectors θ and γ:

$$\min_{u,v,w,\theta,\gamma} \sum_{t=1}^{T} \|Y_\alpha(t) - Y(t)\|^2 \tag{11.6}$$

where Y_α denotes the actual process outputs and T is the total number of time steps in the sequence.

11.3   Unfolding the Network in Time

A common approach to solving the optimization problem as stated in (11.6) is to unfold the recurrent network in time [85]. This practice will convert the representation of the recurrence into a spatial feedforward network. The unfolded network structure is depicted in Figure 11.3, where each time step constitutes one layer of the network. This means that the hidden units at the layer for time step t receive inputs from the preceding layer of time step $t-1$ and then deliver information to the next layer of time step $t+1$. The weight matrices U, V, W, and bias vectors θ and γ are shared in the unfolded network to insure the same dynamics and input-output relations across all time steps.

It bears noting that, owing to constraints of computational resources, the unfolding in practice has to be limited by truncation after a certain number of time steps m. The determination of m is usually based on the heuristics of how many steps to trace back to retrieve adequate information to predict the process outputs $Y(t)$. Unfolding with m steps backwards from each time $t \in \{m+1, T\}$ will produce a set of sliding windows, each of which can be treated as a training sample. Then the task of network learning can be reformulated into the optimization problem of finding optimal weight matrices to minimize the error on all sliding windows outputs.

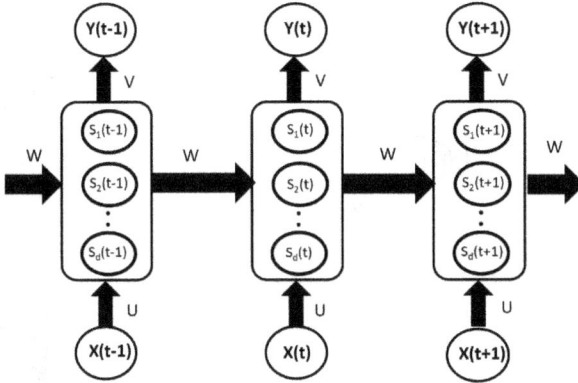

FIGURE 11.3: Unfolded network with shared weights.

11.4 Backpropagation Through Time

As training example we use a sliding window which unfolds the recurrence from time step $t \in m + 1, ..., T - 1, T$ backward till $t - m$ while having outputs only at the final step t. The unfolded network in correspondence to such sliding windows is depicted in Figure 11.4. The error function $E(t)$ for this training example is simply defined as:

$$E(t) = \frac{1}{2} \|Y_\alpha(t) - Y(t)\|^2 \tag{11.7}$$

We then follow a stochastic gradient decent to modify the weight matrices to minimize the error $E(t)$ in (11.7). To this end the partial derivatives of $E(t)$ with respect to weight matrices U, V, and W have to be derived.

We start from the weights V that are used to produce the outputs $Y(t)$. The partial derivatives with respect to V is calculated by

$$\frac{\partial E(t)}{\partial V} = \frac{\partial E(t)}{\partial Y(t)} \cdot \frac{\partial Y(t)}{\partial V} = [Y(t) - Y_\alpha(t)] \oplus S(t) \tag{11.8}$$

where $\oplus$ is the outer product of two vectors.

Regarding the weights W appearing at multiple time steps of the sliding window, we have to sum up their contribution at the each of the steps in calculation of the gradient. Therefore we have

$$\frac{\partial E(t)}{\partial W} = diag\left(\frac{\partial E(t)}{\partial S(t)}\right) \cdot \frac{\partial S(t)}{\partial W} + diag\left(\frac{\partial E(t)}{\partial S(t-1)}\right) \cdot \frac{\partial S(t-1)}{\partial W} + ...$$
$$+ diag\left(\frac{\partial E(t)}{\partial S(t-m+1)}\right) \cdot \frac{\partial S(t-m+1)}{\partial W} \tag{11.9}$$

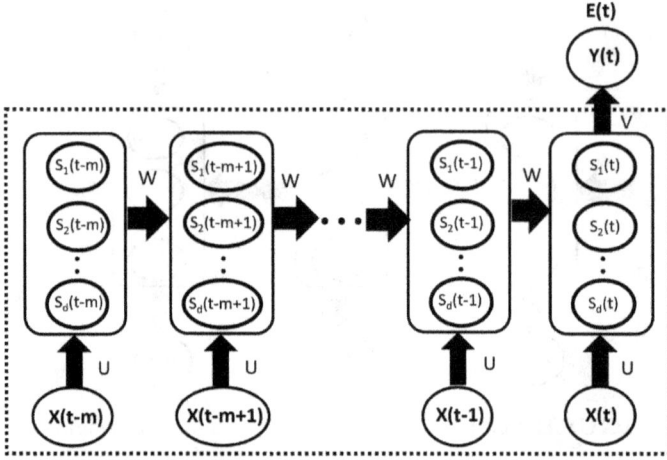

FIGURE 11.4: An unfolded network with output only at the last step.

with $diag\left(\frac{\partial E(\tau)}{\partial S(\tau)}\right)$ indicating the diagonal matrix containing the elements of $\frac{\partial E(\tau)}{\partial S(\tau)}$ for $\tau \in \{t-m+1, ..., t-1, t\}$.

Since the partial derivative of $S(\tau)(\tau = t-m+1, ..., t)$ with respect to W is obtained by

$$\frac{\partial S(t)}{\partial W} = \frac{\partial S(t)}{\partial Z(t)} \cdot \frac{\partial Z(t)}{\partial W} = tanh'[Z(\tau)] \oplus S(\tau-1) \qquad (11.10)$$

where

$$Z(\tau) = \begin{bmatrix} z_1(\tau) \\ \vdots \\ z_n(\tau) \end{bmatrix} = W \cdot S(\tau-1) + U \cdot X(\tau) + \theta \qquad (11.11)$$

$$tanh'(Z(\tau)) = \begin{bmatrix} 1 - tanh^2(z_1(\tau)) \\ \vdots \\ 1 - tanh^2(z_n(\tau)) \end{bmatrix} \qquad (11.12)$$

what remains to be solved (11.9) is to acquire the partial derivatives of $E(t)$ with respect to S at various time steps. These derivatives can be yielded by using a backpropagation in time (BPTT) method as follows.

We start from the final time step t and obtain

$$\frac{\partial E(t)}{\partial S(t)} = \frac{\partial E(t)}{\partial Y(t)} \cdot \frac{\partial Y(t)}{\partial S(t)} = [Y(t) - Y_\alpha(t)]' \cdot V \qquad (11.13)$$

Then we move backwards to the successive proceeding time steps by calculating the partial derivatives $\frac{\partial E(t)}{S(\tau-1)}$ for the time step $\tau - 1$ based on the information available at the step τ. The information propagation is written as follows:

$$\frac{\partial E(t)}{\partial S(\tau - 1)} = \frac{\partial E(t)}{\partial S(\tau)} \cdot \frac{\partial S(\tau)}{\partial S(\tau - 1)} \quad \tau = t - m + 1, ..., t \quad (11.14)$$

$$\frac{\partial S(\tau)}{\partial S(\tau - 1)} = \frac{\partial S(\tau)}{\partial Z(\tau)} \cdot \frac{\partial Z(\tau)}{\partial S(\tau - 1)} = diag[tanh'(Z(\tau)] \cdot W \quad (11.15)$$

where $tanh'(Z(\tau)$ has been defined in (11.12) and $diag[tanh'(Z(\tau)]$ is a diagonal matrix containing the elements of $tanh'(Z(\tau))$.

Likewise, the partial derivatives of E with respect to the weights U are obtained by considering the influences of U at multiple time steps:

$$\frac{\partial E(t)}{\partial U} = diag\left(\frac{\partial E(t)}{\partial S(t)}\right) \cdot \frac{\partial S(t)}{\partial U} + diag\left(\frac{\partial E(t)}{\partial S(t-1)}\right) \cdot \frac{\partial S(t-1)}{\partial U} + ...$$
$$+ diag\left(\frac{\partial E(t)}{\partial S(t-m)}\right) \cdot \frac{\partial S(t-m)}{\partial U}$$
$$(11.16)$$

where the partial derivatives of $S(\tau)(\tau = t - m, ..., t)$ with repect to U are given as

$$\frac{\partial S(\tau)}{\partial U} = \frac{\partial S(\tau)}{\partial Z(\tau)} \cdot \frac{\partial Z(\tau)}{\partial U} \begin{bmatrix} 1 - tanh^2(z_1(\tau)) \\ \vdots \\ 1 - tanh^2(z_n(\tau)) \end{bmatrix} \oplus X(\tau) \quad (11.17)$$

Since $\frac{\partial E(t)}{\partial S(\tau)}$ for $\tau \in \{t - m, ..., t - 1, t\}$ can be yielded according to Eqs. (11.13) to (11.15), we now solve Eq. (11.16) to get the partial gradient concerning U.

Once the full gradient information for a training example (sliding window) is available, the weights will be updated in terms of gradient descent to efficiently reduce the error. This instance-based update will be conducted for all training examples to finish one epoch of learning. Subsequently the updated network will be evaluated on both training and validation data sets to determine if the termination condition is satisfied. In cases of the termination condition not being satisfied, a new epoch of learning will be launched. The complete procedure to learn the weights of recurrent neural networks based on BPTT is formally described in Algorithm 12.

11.5 The Challenge of Long-term Dependencies

One challenge that may arise with the learning of a recurrent network is caused by the long-term dependencies. This is attributed to the fact that an unfolded

---

Algorithm 12 Learning of recurrent neural networks based on BPTT

Require: A set of sliding windows as training examples. The number of hidden units in the network. The learning rate η

Ensure: The connection weights U, V, W of the network

1: Initialize the connection weights U, V, W with random small values

2: **while** termination condition is not met **do**

3:   **for** each sliding window $\langle X_i(t-m), ..., X_i(t) \rangle$ **do**

4:     Forward calculation of the unfolded network output $Y(t)$ based on the inputs $\langle X_i(t-m), ..., X_i(t) \rangle$

5:     Calculating the partial gradient with respect to V

6:     $\frac{\partial E(t)}{\partial V} = [Y(t) - Y_\alpha^i(t)] \oplus S(t)$

7:     Backpropagation of $\frac{\partial E(t)}{\partial S(\tau)}$ with the start from $\tau = t$ as follows

8:     $\frac{\partial E(t)}{\partial S(t)} = [Y(t) - Y_\alpha^i(t)]' \cdot V$

9:     $\frac{\partial E(t)}{\partial S(\tau-1)} = \frac{\partial E(t)}{\partial S(\tau)} \cdot diag[tanh'(Z(\tau)] \cdot W \quad \tau = t-m+1, ..., t$

10:    Calculating the partial gradient with respect to W

11:    $\frac{\partial E(t)}{\partial W} = \sum_{t-m+1}^{t} diag\left(\frac{\partial E(t)}{\partial S(\tau)}\right) \cdot \frac{\partial S(\tau)}{\partial W}$

12:    $\frac{\partial S(\tau)}{\partial W} = tanh'[Z(\tau)] \oplus S(\tau-1) \quad \tau = t-m+1, ..., t$

13:    Calculating the partial gradient with respect to U

14:    $\frac{\partial E(t)}{\partial U} = \sum_{t-m}^{t} diag\left(\frac{\partial E(t)}{\partial S(\tau)}\right) \cdot \frac{\partial S(\tau)}{\partial U}$

15:    $\frac{\partial S(\tau)}{\partial U} = tanh'[Z(\tau)] \oplus X_i(\tau) \quad \tau = t-m, ..., t$

16:    Update the weights based on the gradient information

17:    $W = W + \eta \cdot \frac{\partial E(t)}{\partial W}, U = U + \eta \cdot \frac{\partial E(t)}{\partial U}, V = V + \eta \cdot \frac{\partial E(t)}{\partial V}$

18:   Evaluate the updated network on the training and validation data

---

network is a deep network comprising of a number of layers corresponding to the time steps. Backpropagating the gradient information through many layers may tend to produce partial derivatives that are close to zero, giving rise to the so called vanishing gradient problem. The vanishing gradient will make it hard for the search algorithm to identify the direction to change the weights to quickly reduce the cost function. A detailed explanation of the vanishing gradient with recurrent networks is given in the following. Deeper analysis and treatment of the long-term dependencies can be from [29][13] and [107].

According to the BPTT method described in Section 11.4, the vectors of partial derivatives $\frac{\partial E(t)}{\partial S(\tau)}$ are calculated from time step $\tau = t$ and then will be backpropagated in time. The backpropagation chain is shown in Figure 11.5, where the transition between two consecutive time steps is achieved by the multiplication of a transition matrix $\frac{\partial S(\tau)}{\partial S(\tau-1)}$. It follows that the vector of partial derivatives at time step $t - k, (k \leq m)$ will be obtained by using a chain of matrix multiplications as follows:

$$\frac{\partial E(t)}{\partial S(t-k)} = \frac{\partial E(t)}{\partial S(t)} \cdot \frac{\partial S(t)}{\partial S(t-1)} \cdot \frac{\partial S(t-1)}{\partial S(t-2)} \cdot \ldots \cdot \frac{\partial S(t-k+1)}{\partial S(t-k)}$$

$$= \frac{\partial E(t)}{\partial S(t)} \cdot \prod_{t-k+1}^{t} \frac{\partial S(\tau)}{\partial S(\tau-1)} \qquad (11.18)$$

Based on (11.15) the transition matrices can be reformulated as

$$\frac{\partial S(\tau)}{\partial S(\tau-1)} = \begin{bmatrix} tanh'(z_1(\tau)) & \cdots & 0 \\ \vdots & \ddots & \vdots \\ 0 & \cdots & tanh'(z_n(\tau)) \end{bmatrix} \cdot W \qquad (11.19)$$

Because the derivatives of the *tanh* function are bounded by one and appear as zero at both ends (as seen from Figure 11.6), the transition matrices in (11.19) actually represent a decay of the weights W. Consequently, with small values in the matrices and multiple matrix multiplications (k times in this case), the gradient may decline exponentially fast and finally vanish if k is large. This causes the difficulty of learning the long dependency using the BPTT algorithm since the gradient contribution from remote time steps is likely to be zero.

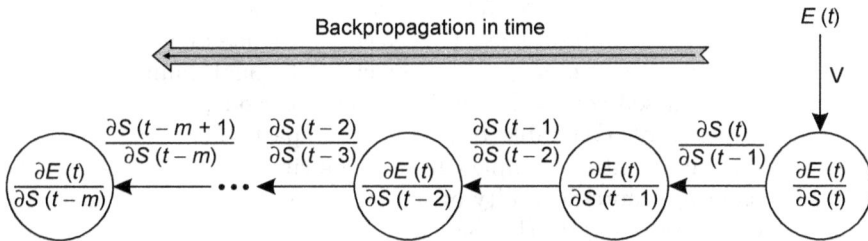

FIGURE 11.5: Calculating partial derivatives in a backpropagation chain.

The risk of the vanishing gradient can be seen more clearly when the transition matrix is identical across the various time steps. We use W' to denote this transition matrix. Suppose W' has an eigendecomposition $W' = Qdiag(\lambda)Q^{-1}$ with Q being orthogonal. The partial derivatives $\frac{\partial E(t)}{\partial S(t-k)}$ at time step $t-k$ becomes

$$\frac{\partial E(t)}{\partial S(t-k)} = \frac{\partial E(t)}{\partial S(t)} \cdot (W')^k = \frac{\partial E(t)}{\partial S(t)} \cdot Qdiag(\lambda)^k Q^{-1} \qquad (11.20)$$

The eigenvalues are raised to the power of k, resulting in those with the original magnitudes less than one to decay to zero. As $\frac{\partial E(t)}{\partial S(t-k)}$ is scaled to $diag(\lambda)^k$, some of its components will eventually be discarded leading to a vanishing gradient.

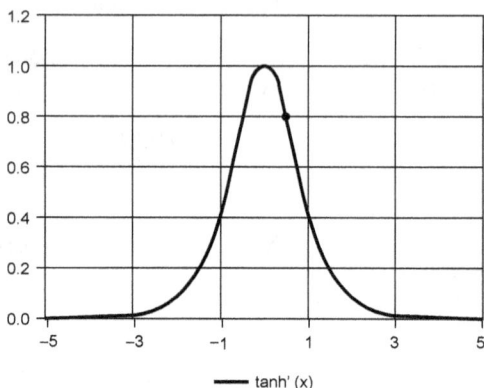

FIGURE 11.6: The derivative of the tanh function [1].

11.6   Long-Short Term Memory (LSTM)

Given that gradients may vanish through a number of time steps, one mitigation approach proposed by Lin et al. [94] was to add skip connections through multiple time steps, based on the idea of building recurrent networks with longer delays [12]. It incorporated direct connections of units from distant past to present to better capture the long-term dependences.

Implementing hidden units as leaky units [101] presents an alternative way to obtain the similar effect of skipping connections through multiple steps. A leaky unit has linear self-connection with a weight near one, so that it enables the direct influence of a variable from a distant past. The time constants of leaky units can be either sampled from a distribution or determined by learning. This brings more flexibility of smooth adjustment than by changing the integer-valued skip length. Having leaky units of different time scales was shown to benefit the learning of long dependency [107].

The LSTM model [71] resembles leaky units by using a self-loop to allow for information to be kept for a longer duration. However, the weights of self-loop in LSTM can change across various time steps instead of remaining constant as in leaky units. The basic idea is that these parameters should be situation dependent and hence they are decided by the gating of some other hidden units. By learning to adapt the self-loop parameters through time, it will be possible to dynamically change the time scales of paths in the model so that a piece information can be memorized or discarded based on the situation in terms of its usefulness.

The block diagram of an LSTM cell is depicted in Figure 11.7, in which the "forget gate" plays a central role by setting the parameter of the linear self-loop of the state of the cell. Moreover, the cell contains the "input gate" and "output gate" to control the input and output flows of information. All these three gates are implemented as hidden units which receive the current external $X(t)$ and preceding outputs of the cells $S(t-1)$ and then produce gating values according to the sigmoid function. LSTM cells can be used to replace the usual hidden units of conventional recurrent neural networks for construction of LSTM recurrent networks.

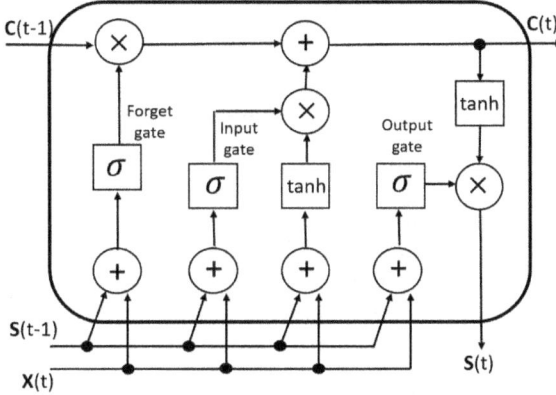

FIGURE 11.7: An LSTM recurrent network cell.

More concretely, the state of a cell i has linear self-connection as similar to leaky units, but its parameter is set by the gating value of the forget gate as follows:

$$f_i(t) = \sigma[b^f(i) + U^f(i,:) \cdot X(t) + W^f(i,:) \cdot S(t-1)] \tag{11.21}$$

where b^f, U^f, W^f are respectively the bias vector, input weight matrix, and recurrent weight matrix of the forget gates. Likewise, the input gate uses its own parameters from the bias vector b^g, input weight matrix U^g and recurrent weight matrix W^g respectively to produce the value of input gating as

$$g_i(t) = \sigma[b^g(i) + U^g(i,:) \cdot X(t) + W^g(i,:) \cdot S(t-1)] \tag{11.22}$$

Thus, the state of the LSTM cell can now be updated using both the self-loop parameter and the input gating value as

$$C_i(t) = f_i(t) \cdot C_i(t-1) + g_i(t) \cdot \sigma[b(i) + U(i,:) \cdot X(t) + W(i,:) \cdot S(t-1)] \tag{11.23}$$

where b, U, and W refers to the bias vector, input weight matrix, and recurrent weight matrix for the cells of the LSTM model.

The output of the cell at time t, $Si(t)$, is decided by both the cell state and the value of the output gate:

$$S_i(t) = tanh(C_i(t)) \cdot q_i(t) \tag{11.24}$$

$$q_i(t) = \sigma[b^o(i) + U^o(i,:) \cdot X(t) + W^o(i,:) \cdot S(t-1)] \tag{11.25}$$

where b^o, U^o, W^o denote respectively the bias vector, input weight matrix, and recurrent weight matrix of the output gates.

Finally, recent research has demonstrated the powerful performance of LSTM networks in learning the long-term dependency based on complex sequential information, particularly in the applications of machine translation [130], speech recognition [64], image captioning [17], time series [144][146] as well as industrial process modeling [158].

11.7 Other Recurrent Networks

This section briefly discusses the other three types of structures of recurrent networks that have not been addressed in the preceding sections. They include recurrent networks with unfolding outputs at all time steps, gated recurrent networks, as well as echo state networks.

11.7.1 Unfolding outputs at all steps

In Section 11.4 we discussed the recurrent network learning with the unfolded structure having outputs only at the last time step. Alternatively, we can also consider an unfolded network that has outputs at each time step, as shown in Figure 11.8. In that case, the new error function $TE(t-m,...,t-1,t)$ has to be redefined as:

$$TE(t-m,...,t-1,t) = \sum_{\tau=t-m}^{t} E(\tau) \tag{11.26}$$

where

$$E(\tau) = \frac{1}{2}\|Y_\alpha(t) - Y(\tau)\|^2 \quad \tau = t-m,...,t \tag{11.27}$$

$Y_\alpha(\tau)$ denotes the real process outputs at time τ.

Having the new error function in (11.26), the gradients that are needed for learning will be calculated as the sum of the partial derivatives of $E(\tau)$ with respect to weights across all time steps. Fortunately, these partial derivatives can be obtained by using the same method as given in Section 11.4 for calculating the derivatives of $E(t)$. Of course, Algorithm 12 for learning also needs to be revised accordingly by considering the gradients of $TE(t-m,...,t-1,t)$ rather than $E(t)$ in weight updating.

11.7.2 Gated recurrent networks

More recently, gated recurrent networks were proposed [78][18], which differ from LSTM by adopting a single gate, termed as "update gate", to control the integration of the old and new target states in their units. A unit that is used in a gated recurrent network is also known as gated recurrent unit

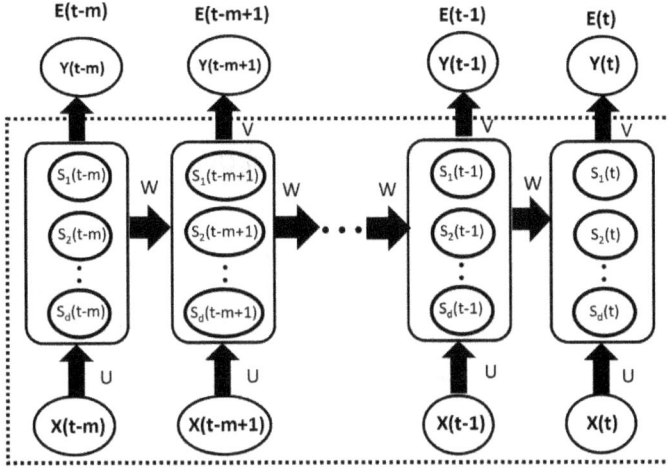

FIGURE 11.8: An unfolded network with outputs at each time step.

(GRU) [20][38][86][99][129]. The state update of GRU in time step t can be formulated as:

$$S_i^t = u_i^{t-1} S_i^{t-1} + (1 - u_i^{t-1})\sigma(b(i) + U(i,:)X(t-1) + W(i,:)[.., r_j^{t-1} S_j^{t-1}, ..]') \tag{11.28}$$

where u_i^{t-1} denotes the value of the update gate, while r_j^{t-1} stands for the value of the "reset gate" which is used to control the degree of involvement of GRU j in computing the new target state. The update and reset gates are defined as sigmoid units with their values being calculated as

$$u_i^t = \sigma(b^u(i) + U^u(i,:) \cdot X(t) + W^u(i,:) \cdot S^t) \tag{11.29}$$

$$r_i^t = \sigma(b^r(i) + U^r(i,:) \cdot X(t) + W^r(i,:) \cdot S^t) \tag{11.30}$$

where b^u, U^u, W^u denote respectively the bias vector, input weight matrix, and recurrent weight matrix of the update gates, while b^r, U^r, W^r refer to the corresponding weights of the reset gates.

11.7.3 Echo state networks

To avoid the difficulty of parameter learning with a large recurrent network, echo state networks (ESNs) [74][75] were proposed with the idea of only learning the weights of the output layer while making the input and recurrent weights nontrainable. ESNs are considered as one class of networks in reservoir computing, which aims to utilize the temporal features derived from hidden units to capture the history of past inputs. The other class of networks belonging to reservoir computing are liquid state machines [95], which differ from ESNs in using binary-valued spike neurons rather than hidden units with continuous outputs.

In the design of an echo state network, it is crucial to set the input and recurrent weights to reach the desired property of the dynamic system, as measured based on the Jacobian of the hidden state transitions. One important index of the Jacobian is the spectral radius, which is defined as the maximum of the absolute values of its eigenvalues. The recent research [156] on ESNs suggested having a spectral radius of much more than unity to ensure a resultant large variation when an initial perturbation is propagated through a number of time steps. Note that a large spectral radius will not lead to unbounded dynamics in forward propagation due to the saturating effect of the tanh activation function.

Once the input and recurrent weights are successfully generated, the ESN will be able to capture the essential information of an input sequence, which is transformed into a fixed-length vector in terms of the hidden units. The network output will then be produced based on the values of the hidden units, typically through a linear regression function. A simple learning algorithm [74] can be applied here to identify the output weights in an attempt to minimize the mean squared error.

12

Convolutional Neural Networks (CNN)

12.1 Introduction

Convolutional Neural Networks [89] (CNN), are a category of neural networks that are specially designed to process data represented with a grid structure, such as images and time series data. The word "convolutional" indicates that convolutional networks use the mathematical operation of convolution. The convolution operation is introduced in CNNs to replace the general matrix multiplication in one or more layers of the network.

On the other hand, convolutional networks are quite similar to feedforward multilayer neural networks. They consist of inter-connected neurons, each of which receives multiple inputs from the preceding layer and then produces an output through an activation function. As a whole the total network still behaves as a non-linear and differentiable mapping from input to output spaces. A distinguishing property of CNNs is that they enable incorporating grid structure of data into model architecture such that both neural connections and their weights can be dramatically reduced, which improves computational efficiency in learning and inference of the model as well as saves memory for data (weight) storage.

Convolutional networks have been at the forefront in the history of deep learning. Indeed, LeNet-5 [88], the first example of CNNs was proposed much before deep learning methods received wide acceptance. The CNNs were also among some of the first deep learning models that were successfully trained with back-propagation. This gives strong evidence of the higher computational efficiency of CNNs than regular fully connected networks. The success of CNNs paved the way to the recognition of neural networks in general and promoted fruitful research into other deep learning models and algorithms.

There have been many convolutional networks successfully applied to many problems with practical domains. Examples of early application include the CNN-based OCR and handwriting recognition systems [123], developed by Microsoft. The intense commercial interest in CNNs started when the deep CNN created by [84] achieved record-breaking results with classifying 1.2 million images into 1000 classes. This advance won the ImageNet Large Scale Visual Recognition Challenge (ILSVRC) in 2012 and also resulted in dominating impact on the subsequent events of ILSVRC. A comprehensive review of the models, methodologies and applications of CNNs can be found in [28].

The remainder of this chapter is organized as follows. Section 12.2 gives an overview of the general architecture of CNNs. The two important layers of CNNs, convolutional layer and pooling layer, are described in Sections 12.3 and 12.4 respectively. Section 12.5 briefly discusses the learning issue of CNNs. Finally, Section 12.6 presents recurrent CNNs that aim to handle grid data in time series.

12.2 Architecture Overview

Let's consider images of the size $p \times q \times r$, where p, q, and r denotes the width, height, and number of color channels, respectively. If we use a regular fully connected network to process these images, a neuron in the first hidden layer will have $p \times q \times r$ weights (plus the bias) for being connected to each pixel of the image. As the layer contains multiple neurons, the total number of weights will add up very quickly. This number of parameters would only be manageable when the size of images is small. It will be hard to scale up to large images for which the width and height are above 200 for instance. Moreover, a huge number of parameters in the neural network will increase the risk of overfitting in model training.

To more efficiently deal with large scale inputs, convolutional networks are specially designed with convolutional layers to depict data of grid topology. Each neuron in a convolutional layer is only connected to units of a local region rather than all neurons of the preceding layer. The task of these neurons is to perform a convolution of the pixels of the regions that they are respectively associated with. The results of convolution are then transformed by a non-linear, rectified function to yield local features of the corresponding regions.

The coupling of convolution and non-linearity (via the Rectified Linear Unit function) constitutes one kernel in the convolutional layer. The kernel is useful to detect a certain local feature across the whole range of the data, giving rise to a feature map [87][98][154]. Since we are usually interested in detecting more than a single feature, several groups of neurons have to be arranged in the convolutional layer to implement multiple kernel functions for creating different feature maps. Sometimes such a kernel is also referred to as filter in the literature. A general architecture of convolutional networks is shown in Figure 12.1. It comprises a stack of stages with the purpose of learning feature representations at various abstracting levels. Each stage consists of a convolutional layer and a pooling layer. The convolutional layer, containing convolution and ReLU, acts as kernels to produce multiple feature maps, while the pooling layer is responsible for performing a downsampling operation along the spatial dimensions of the feature maps [114]. At last, the fully connected layers compute the final outputs, i.e., the score of each class.

Finding the best CNN architecture is problem and data dependent. There are many ongoing pieces of research that frequently reveal a new architecture

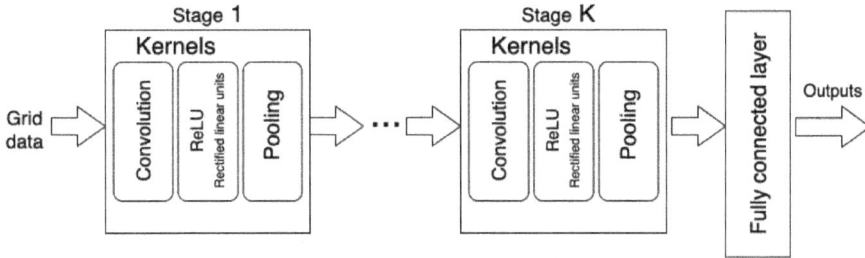

FIGURE 12.1: The general architecture of CNN.

for a given benchmark that improves model performance. But this issue is not going to be addressed here. In the subsequent sections, we will only explain the convolutional and pooling layers that are important building blocks for construction of CNNs.

12.3 Convolutional Layer

Convolutional layer has been designed based on two main ideas: sparse connectivity and parameter sharing [154], which is further explained in the sequel:

Sparse connectivity. As is noted before, a neuron in the convolutional layer aims to detect a feature in a local area and hence it is only connected to a small part of units in the preceding layer. This method is very advantageous in resulting in fewer parameters of the model as well as fewer operations to calculate model outputs, in comparison to an ordinary fully connected network. Suppose the images to process are of the size $p \times q \times l$ (with p, q being the width and height, respectively). If we use a network of full connection, the first hidden layer with s neurons will have a total of p×q×s connections and parameters, and the runtime to calculate the hidden layer outputs (via matrix multiplication) will be $O(p \times q \times s)$. Nevertheless, a neuron in the first convolutional layer of CNN is merely connected to a local region of k pixels instead of the entire image. It follows that a feature map (with s neurons) in the convolutional layer will have k×s connections and thereby the runtime will be $O(k \times s)$. Since k is usually several orders of magnitude smaller than p×q, we significantly reduce the complexity and computing expense with this sparse connectivity.

Parameter sharing. It refers to the use of a weight by multiple connections of the network rather than assigning a unique weight to each neural connection. Given that a feature map in the convolutional layer is tasked to identify a single feature across the input space, all its neurons are arranged to have an identical set of parameters to ensure the same kernel function be applied at different locations. In other words, the value of a weight employed at one position is tied to the value of a weight employed at any other locations.

Therefore, only k parameters (plus the bias) are needed by the feature map despite the k×s connections. This indicates a dramatic reduction of model parameters, leading to a much lower memory requirement of the model and higher statistical efficiency in learning.

The rule of parameter sharing also brings the property that the convolutional layer is of **equivariance** to translation of input. This can be explained by the fact that shared parameters cause the same kernel function to be applied everywhere such that the convolution of translated input is equal to the result of translating the convolution of the original input. Arguably, when an image shifts, its new feature map can be obtained by performing the same shift to the feature map of the original image. Hence no information will get lost in the convolutional layer with occurrences of input data shifting.

In a convolutional stage, a kernel is applied to different local regions by sliding across the entire range of the input data. The number of pixels the kernel moves at a time is called stride. At each local region, the kernel first calculates the sum of the products of each kernel weight and the corresponding input element, which is then transformed by the Rectified Linear Unit (ReLU) function to produce the output of the connected neuron in the convolutional layer. The ReLU function used here is given by:

$$f(x) = \begin{cases} x, & x \geq 0 \\ 0, & x < 0 \end{cases} \tag{12.1}$$

Figure 12.2 gives an example on how a feature map is generated by performing convolution to an image of the size 4×4×1. The local regions that the kernel is applied to, has the size of 3×3×1, and the stride is set to be one. The resultant 2-D feature map provides information about the values of the feature at different locations.

12.4   Pooling Layer

The pooling layer is designed for downsampling of feature maps in order to reduce the number of outputs from the convolutional layer. The pooling function constructs the layer by summarizing rectangular neighborhoods in the feature maps with statistical values. Two commonly used pooling methods are max-pooling and average pooling [117][90], as described in the following:

- Max-pooling: it reports the maximum value of a rectangular neighborhood. We slide the neighborhood (also termed as pool) over the entire feature map to obtain the maximum outputs of the convolutional layer at different locations.

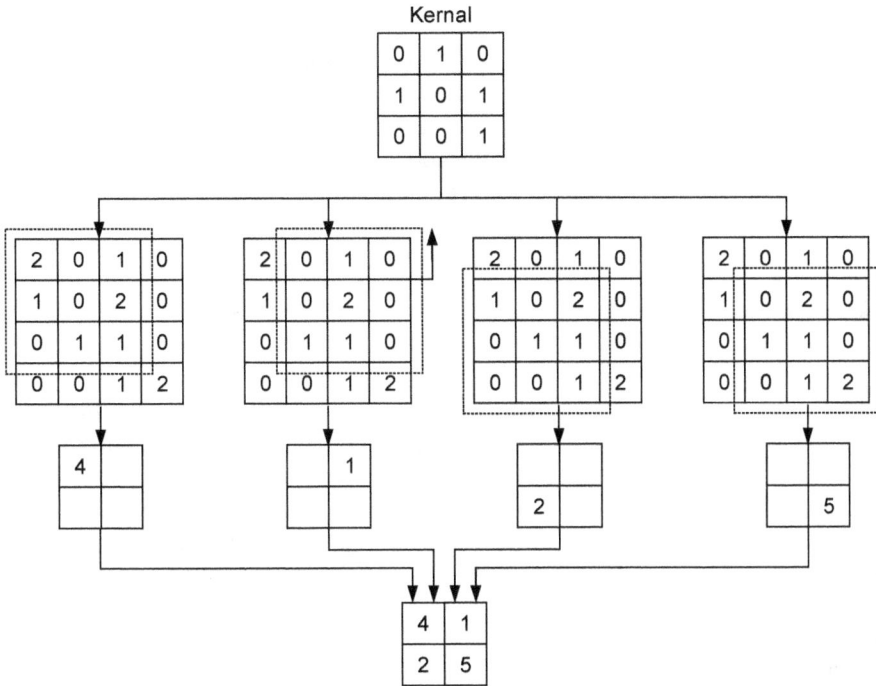

FIGURE 12.2: Generation of a feature map via convolution.

- Average pooling: it calculates the average of the values of a rectangular neighborhood. We slide the pool over the entire feature map so as to obtain the mean of outputs of the convolutional layer at different locations.

An example of the pooling operations is given in Figure 12.3, in which the max-pooling and average-pooling are applied to a feature map of the size 8×8. A 4×4 pool slides across the dimensions of the feature map with the stride of two. Both pooling operations lead to compression of the 16 outputs in the feature map into four units of the pooling layer.

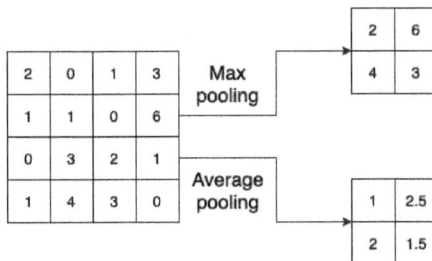

FIGURE 12.3: Example of pooling applied to a feature map.

It is worth noting that the pooling layer representation is approximately invariant to small translation of input data. This is because each unit in the layer provides a summary statistic of a neighborhood so that it is quite insensitive to slight changes in the pool. Particularly, with max-pooling, the output of a pooling unit will remain unchanged as long as the maximum output of the corresponding neighborhood does not change. This invariance property is beneficial for the feature detection purpose in the sense that the identified feature values will receive little influence from input data shifts.

12.5   Learning of CNN

The learning of CNN is to adapt the weights of the network to minimize the differences between target labels and predicted outputs, as measured by a loss function E. The most commonly used method is gradient descent, which revises the weights in terms of negative gradient in order to quickly reduce the function E. Generally, the backpropagation algorithm can still be employed here to acquire the gradient and update the weights iteratively.

However, special treatments have to be made for learning with a convolutional layer, given that neurons in this layer share their weights as if they belong to the same feature map. Let's consider a weight $w_{m,n}^l$ of connections in a convolutional layer l. Because this weight is used in calculating outputs of all neurons of the associated feature map, its influence to the loss E has to add up for the whole feature map. In view of this, the partial derivative of E with respect to $w_{m,n}^l$ is expressed as:

$$\frac{\partial E}{\partial w_{m,n}^l} = \Sigma_i \Sigma_j \frac{\partial E}{\partial \alpha_{i,j}^l} \cdot \frac{\partial \alpha_{i,j}^l}{\partial w_{m,n}^l} \tag{12.2}$$

Here $\alpha_{i,j}^l$ denotes the linear combination of inputs for a unit indexed by (i, j) in the feature map. As inputs to the feature map are actually outputs from the preceding layer l-1, we can formulate $\alpha_{i,j}^l$ as

$$\alpha_{i,j}^l = \Sigma_u \Sigma_v O_{i.s+u,j.s+v}^{l-1} w_{u,v}^l + b^l \tag{12.3}$$

where $O_{i.s+u,j.s+v}^{l-1}$ is an output of the local region in the preceding layer that is connected to the unit (i, j) in the next layer, b^l is the bias, and s is the stride of convolution. Based on 12.3, we obtain the following:

$$\frac{\partial \alpha_{i,j}^l}{\partial w_{m,n}^l} = \frac{\partial \left(\Sigma_u \Sigma_v O_{i.s+u,j.s+v}^{l-1} w_{u,v}^l + b^l \right)}{\partial w_{m,n}^l} = O_{i.s+m,j.s+n}^{l-1} \tag{12.4}$$

Further we define the error term of a unit as

$$\frac{\partial E}{\partial \alpha_{i,j}^l} = \delta_{i,j}^l \tag{12.5}$$

Consequently, the derivative of E with respect to $w_{m,n}^l$ can be rewritten as

$$\frac{\partial E}{\partial w_{i,j}^l} = \Sigma_i \Sigma_j \delta_{i,j}^l . O_{i.s+m,j.s+n}^{l-1} \tag{12.6}$$

In principle, the error terms can be acquired backwards as specified in the backpropagation algorithm. But, note that a pooling layer has no weights. This entails the following extra rules to be honored in backpropagation when routing gradient from the pooling layer to its preceding convolutional layer.

Backpropagation from a max-pooling layer: we assign the error term to the unit in the pool which has the maximum output, while all other units in the pool get zero error (term) since they have no contribution to the pooling unit. For doing this, the wining unit that has the maximum output from the pool needs to be recorded already in the forward pass.

Backpropagation from an average-pooling layer: all units in the pool receive an equal error term, which is the error term of the pooling unit divided by the number of units of the pool. The reason of this assignment is that all units in the pool have equal contribution to the pooling unit in the next layer.

In many situations, designing a good CNN architecture and training the network from scratch is highly expensive and time consuming. One practical way to relieve the learning burden is to perform transfer learning based on a pre-trained CNN that has been optimized for a similar task. The pre-trained network can be further updated and refined using the new data of the underlying task. Sometimes we can even freeze several layers and only adjust parameters of the input and output layers of the pre-trained network. The method of transfer learning is also very helpful when there is a lack of sufficient data to train a complex CNN.

There are already a few CNN models which can be reused for adaptation to solve new problems. The most well known CNN architectures include LeNet-5 [88], VGG [124], AlexNet [84], GoogLeNet [131], ResNet [67], etc. Caffe [77], as a modifiable framework of the state-of-the-art deep learning methods, provides a collection of reference models of general-purpose CNNs that can be selected/adopted for transfer learning to tackle a new task at hand.

12.6   Recurrent CNN

In CNN networks, information flows in a forward path, and consequently, CNNs are incapable of capturing contextual dependencies [144]. To solve this problem, [92] proposed the recurrent convolutional layers (RCL), which replaces the convolutional layers by incorporating the recurrent connections into each convolutional layer. Therefore, RCL is the core of RCNN. The recurrent connections expand the network depth while the number of parameters is kept constant by weight sharing. A general RNN network is defined as follows:

$$a(t) = g(w_f x(t) + w_r a(t-1) + b_a) \tag{12.7}$$

where x and a indicate the feed-forward and recurrent input (hidden state), respectively, w_f and w_r indicate feed-forward and recurrent weights, respectively, and b_a is the bias.

As mentioned before, in RCL layers, recurrent connections are incorporated into the convolutional layer. Therefore, according to 12.7 its net input is given by

$$Z_{ijk}(t) = (w_f^k)^T x^{(i,j)}(t) + (w_r^k)^T a^{(i,j)}(t-1) + b_a \tag{12.8}$$

where (i, j) indicates the center of the vectorized square patch of the Kth feature map, $x^{(i,j)}$ indicates the feed-forward input and it is the vectorized square patch of the feature map from the previous layer, $a^{(i,j)}$ indicates the recurrent input and it is the vectorized square patch of the feature map from the current layer, w_f^k, w_r^k and b_a indicates the feed-forward weights, recurrent weights and, the bias for the Kth feature map respectively. The first term of 12.8 is calculated using standard CNN and the second term is calculated using the recurrent connections and both of them take the form of convolution. The hidden state of the RCL unit is given by:

$$a_{ijk}(t) = f(h(Z_{ijk}(t))) \tag{12.9}$$

where h is the activation function and the Relu function is usually used:

$$h(Z_{ijk}(t)) = max(0, Z_{ijk}(t)) \tag{12.10}$$

and f is local response normalization (LRN):

$$f(h_{ijk}(t)) = \frac{h_{ijk}(t)}{\left(1 + \frac{\alpha}{M} \sum_{k'=max(0,k-\frac{M}{2})}^{min(K,k+\frac{M}{2})} (h_{ijk'})^2\right)^{\beta}} \tag{12.11}$$

where α and β are constants that control the amplitude of LRN, K is the number of feature maps in the current layer, M is the neighborhood length for normalization, and k is the ith filter output. LRN is a non-trainable layer that simulate the lateral inhibition in the cortex, where different features compete for high activities.

If we unfold an RCL layer for T time steps, a feed-forward subnetwork of depth T+1 is obtained Figure 12.4(a), Figure 12.4(b). T is a hyperparameter, and the effective receptive field of each unit expands when T (iteration number) is bigger [92][93][136]. Therefore, the RCL can capture more contextual information.

RCNN comprises a stack of RCL units with feed-forward connections between neighbors. There are two structures to develop an RCNN [93]:

- The first approach unfolds the RCLs individually for T time steps and then feeding to the next RCL Figure 12.4(a). This approach multiplicatively expands the depth of the network.

- In the second approach, at each time step, the states of all RCLs are updated consecutively and the output of every neuron in a current layer connects to the input of every neuron in the next layer Figure 12.4(b). This unfolding approach additively expands the depth of the network.

If we have a L RCL unit and T time step, then the largest depth of the network is L×T and N+T in the first and second approaches, respectively. The second approach is more computationally expensive because the feed-forward input needs to be updated at each time step, but the first approach needs to be updated once. Additionally, the effective RF is wider in the first approach.

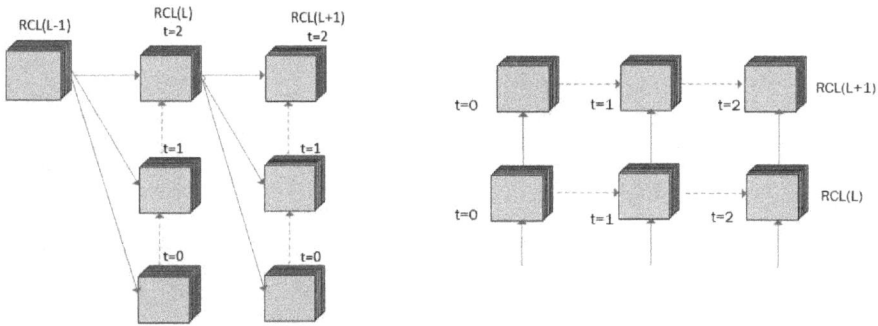

FIGURE 12.4: (a) Multiplicatively unfold two RCLs with T=2 (b) Additively unfold two RCLs with T=2.

The training process is performed by backpropagation through time (BPTT), in which all the RCNNs unfolded to feed-forward networks, and the BP algorithm is applied.

Bibliography

[1] keisan.casio.com. urlhttps://keisan.casio.com/exec/ystem/15411343272927. Accessed: 2021-09-12.

[2] Aghajary, M.M. and A. Gharehbaghi. A novel adaptive control design method for stochastic nonlinear systems using neural network. *Neural Computing and Applications*, 33: 9259–9287, 2021.

[3] Ahlstrom, C., K. Höglund, P. Hult, J. Häggström, C. Kvart and P. Ask. Assessing aortic stenosis using sample entropy of the phonocardiographic signal in dogs. *IEEE Trans. Biomed. Eng.*, 55: 2107–2109, 2008.

[4] Anderson, J.A. *An Introduction to Neural Networks*. MIT Press, Boston, 1995.

[5] Anguera, X., R. Macrae and N. Oliver. Partial sequence matching using an unbounded dynamic time warping algorithm. In *Acoustics Speech and Signal Processing (ICASSP), 2010 IEEE International Conference on*, pp. 3582–3585, March 2010.

[6] Arslan, L.M. and J.H.L. Hansen. Selective training for hidden markov models with applications to speech classification. *IEEE Transactions on Speech and Audio Processing*, 7(1): 46–54, 1999.

[7] Baum, L.E. and T. Petrie. Statistical inference for probabilistic functions of finite state markov chains. *Ann. Math. Statist.*, 37(6): 1554–1563, 12 1966.

[8] Baum, L.E., T. Petrie, G. Soules and N. Weiss. A maximization technique occurring in the statistical analysis of probabilistic functions of markov chains. *Ann. Math. Statist.*, 41: 164–171, 1970.

[9] Begg, R.K., M. Palaniswami and B. Owen. Support vector machines for automated gait classification. *IEEE Transactions on Biomedical Engineering*, 52: 828–838, 2005.

[10] Belhumeur, P.N., J.P. Hespanha and D.J. Kriegman. Eigenfaces vs. fisherfaces: Recognition using class specific linear projection. *IEEE Trans. Pattern Analyse and Machine Intel.*, 19: 711–720, 1997.

[11] Bengio, Y., Renato de Mori, G. Flammia and R. Kompe. Global optimization of a neural network-hidden markov model hybrid. *Neural Networks, IEEE Transactions on*, 3(2): 252–259, 1992.

[12] Bengio, Y. Artificial neural networks and their application to sequence recognition. 1993.

[13] Bengio, Y., P. Simard and P. Frasconi. Learning long-term dependencies with gradient descent is difficult. *IEEE Transactions on Neural Networks*, 5(2): 157–166, 1994.

[14] Bourlard, H. and C.J. Wellekens. Speech pattern discrimination and multilayer perceptrons. *Comput., Speech & Language*, 3: 1–19, 1989.

[15] Bourlard, H. and C.J. Wellekens. Links between markov models and multilayer perceptrons. *IEEE Trans. Pat. Anal. and Mach. Intel.*, 12: 1167–1178, 1990.

[16] Chen, B. and P. Willett. Detection of hidden markov model transient signals. *IEEE Transactions on Aerospace and Electronic Systems*, 36(4): 1253–1268, 2000.

[17] Chen, M., G. Ding, S. Zhao, H. Chen, Q. Liu and J. Han. Reference based lstm for image captioning. In *Thirty-first AAAI Conference on Artificial Intelligence*, 2017.

[18] Cho, K., B. Van Merriënboer, D. Bahdanau and Y. Bengio. On the properties of neural machine translation: Encoder-decoder approaches. *arXiv preprint arXiv:1409.1259*, 2014.

[19] Christiansen, R. and C. Rushforth. Detecting and locating key words in continuous speech using linear predictive coding. *IEEE Transactions on Acoustics, Speech, and Signal Processing*, 25(5): 361–367, 1977.

[20] Chung, J., C. Gulcehre, K. Cho and Y. Bengio. Empirical evaluation of gated recurrent neural networks on sequence modeling. *arXiv preprint arXiv:1412.3555*, 2014.

[21] Ciaccio, E.J., S.M. Dunn and M. Akay. Biosignal pattern recognition and interpretation systems. 3. Methods of classification. *Engineering in Medicine and Biology Magazine, IEEE*, 13(1): 129–135, 1994.

[22] Cohen, A. *Biomedical Signal Processing*. CRC Press, Florida, 1988.

[23] Cortes, C. and V.N. Vapnik. Support vector networks. *Mach. Learn.*, 20: 273–297, 1995.

[24] Cortes, C. and V. Vapnik. Support-vector networks. *Machine Learning*, 20(3): 273–297, 1995.

[25] Crouse, M.S., R.D. Nowak and R.G. Baraniuk. Wavelet-based statistical signal processing using hidden markov models. *IEEE Transactions on Signal Processing*, 46(4): 886–902, 1998.

[26] Davis, S. and P. Mermelstein. Comparison of parametric representations for monosyllabic word recognition in continuously spoken sentences. *IEEE Transactions on Acoustics, Speech, and Signal Processing*, 28(4): 357–366, 1980.

[27] de Vos, J.P. and M.M. Blanckenberg. Automated pediatric cardiac auscultation. *IEEE Trans. Biomed. Eng.*, 54: 244–252, 2007.

[28] Dhillon, A. and G.K. Verma. Convolutional neural network: A review of models, methodologies and applications to object detection. *Progress in Artificial Intelligence*, 9(2): 85–112, 2020.

[29] Doya, K. Bifurcations of recurrent neural networks in gradient descent learning. *IEEE Transactions on Neural Networks*, 1(75): 218, 1993.

[30] Du, J., C. Gerdtman, A. Gharehbaghi and M. Lindén. A signal processing algorithm for improving the performance of a gyroscopic head-borne computer mouse. *Biomedical Signal Processing and Control*, 35: 30–37, 2017.

[31] Dugast, C., L. Devillers and X. Aubert. Combining tdnn and hmm in a hybrid system for improved continuous-speech recognition. *IEEE Trans. Speech and Audio Processing*, 2: 217–223, 1994.

[32] Dutoit, T. and F. Marques. *Applied Signal Processing*. Springer, New York, 2009.

[33] Ebrahimi, Z., M. Loni, M. Daneshtalab and A. Gharehbaghi. A review on deep learning methods for ecg arrhythmia classification. *Expert Systems with Applications: X*, 7: 100033, 2020.

[34] Elman, J.L. Finding structure in time. *Cognitive Science*, 14(2): 179–211, 1990.

[35] Feng, L., X. Zhao, Y. Liu, Y. Yao and B. Jin. A similarity measure of jumping dynamic time warping. In *Fuzzy Systems and Knowledge Discovery (FSKD), 2010 Seventh International Conference on*, volume 4, pp. 1677–1681, Aug 2010.

[36] Forney, G.D. The viterbi algorithm. *Proceedings of the IEEE*, 61(3): 268–278, 1973.

[37] Furui, S. Cepstral analysis technique for automatic speaker verification. *IEEE Transactions on Acoustics, Speech, and Signal Processing*, 29(2): 254–272, 1981.

[38] Gao, S., Y. Zheng and X. Guo. Gated recurrent unit-based heart sound analysis for heart failure screening. *Biomedical Engineering Online*, 19(1): 1–17, 2020.

[39] Garcia-Moral, A.I., R. Solera-Urena, C. Pelaez-Moreno and F. Diaz-de Maria. Data balancing for efficient training of hybrid ann/hmm automatic speech recognition systems. *IEEE Transactions on Audio, Speech, and Language Processing*, 19(3): 468–481, 2011.

[40] Gharehbaghi, A., P. Ask and A. Babic. A pattern recognition framework for detecting dynamic changes on cyclic time series. *Pattern Recognition*, 48(3): 696–708, 2015.

[41] Gharehbaghi, A., P. Ask, M. Lindén and A. Babic. A novel model for screening aortic stenosis using phonocardiogram. pp. 48–51. *In*: Henrik Mindedal and Mikael Persson (eds.). *16th Nordic-Baltic Conference on Biomedical Engineering, volume 48 of IFMBE Proceedings*. Springer International Publishing, 2015.

[42] Gharehbaghi, A., P. Ask, E. Nylander, B. Janerot-Sjoberg, I. Ekman, M. Linden and A. Babic. A hybrid model for diagnosing sever aortic stenosis in asymptomatic patients using phonocardiogram. In *World Congress on Medical Physics and Biomedical Engineering, June 7–12, 2015, Toronto, Canada*, volume 51 of IFMBE Proceedings, pp. 1006–1009. Springer International Publishing, 2015.

[43] Gharehbaghi, A. and A. Babic. Structural risk evaluation of a deep neural network and a markov model in extracting medical information from phonocardiography. *Studies in Health Technology and Informatics*, 251: 157–160, 2018.

[44] Gharehbaghi, A. and A. Babic. A-test method for quantifying structural risk and learning capacity of supervised machine learning methods. In *Studies in Health Technology and Informatics*, volume 289, pp. 132–135, 2022.

[45] Gharehbaghi, A., A. Babic and A.A. Sepehri. A machine learning method for screening children with patent ductus arteriosus using intelligent phonocardiography. In *EAI International Conference on IoT Technologies for HealthCare*, pp. 89–95. Springer, 2018.

[46] Gharehbaghi, A., A. Babic and A.A. Sepehri. Extraction of diagnostic information from phonocardiographic signal using time-growing neural network. pp. 849–853. *In*: Lenka Lhotska, Lucie Sukupova, Igor Lacković, and Geoffrey S. Ibbott (eds.). World Congress on Medical Physics and Biomedical Engineering 2018. Singapore, 2019. Springer Singapore.

[47] Gharehbaghi, A., M. Borga, B.J. Janerot-Sjöberg and P. Ask. A novel method for discrimination between innocent and pathological heart murmurs. *Medical Engineering and Physics*, 37(7): 674–682, 2015.

[48] Gharehbaghi, A., T. Dutoit, P. Ask and L. Sörnmo. Detection of systolic ejection click using time growing neural network. *Medical Engineering and Physics*, 36(4): 477–483, 2014.

[49] Gharehbaghi, A., T. Dutoit, A. Sepehri, P. Hult and P. Ask. An automatic tool for pediatric heart sounds segmentation. In *2011 Computing in Cardiology*, pp. 37–40, 2011.

[50] Gharehbaghi, A., T. Dutoit, A.A. Sepehri, A. Kocharian and M. Lindén. A novel method for screening children with isolated bicuspid aortic valve. *Cardiovascular Engineering and Technology*, 6(4): 546–556, 2015.

[51] Gharehbaghi, A., I. Ekman, P. Ask, E. Nylander and B. Janerot-Sjoberg. Assessment of aortic valve stenosis severity using intelligent phonocardiography. *International Journal of Cardiology*, 198: 58–60, 2015.

[52] Gharehbaghi, A. and M. Lindén. *An Internet-Based Tool for Pediatric Cardiac Disease Diagnosis Using Intelligent Phonocardiography*, pp. 443–447. Springer International Publishing, 2016.

[53] Gharehbaghi, A. and M. Lindén. A deep machine learning method for classifying cyclic time series of biological signals using time-growing neural network. *IEEE Transactions on Neural Networks and Learning Systems*, 29(9): 4102–4115, 2017.

[54] Gharehbaghi, A., M. Lindén and A. Babic. A decision support system for cardiac disease diagnosis based on machine learning methods. *Stud. Health Technol. Inform.*, 235: 43–47, 2017.

[55] Gharehbaghi, A., M. Lindén and A. Babic. An artificial intelligent-based model for detecting systolic pathological patterns of phonocardiogram based on time-growing neural network. *Applied Soft Computing*, 83: 105615, 2019.

[56] Gharehbaghi, A. and M. Lindén. A deep machine learning method for classifying cyclic time series of biological signals using time-growing neural network. *IEEE Transactions on Neural Networks and Learning Systems*, 29(9): 4102–4115, 2018.

[57] Gharehbaghi, A., A.A. Sepehri and A. Babic. Distinguishing aortic stenosis from bicuspid aortic valve in children using intelligent phonocardiography. In *8th European Medical and Biological Engineering Conference, EMBEC 2020, 29 November 2020 through 3 December 2020*, pp. 399–406. Springer Science and Business Media Deutschland GmbH, 2021.

[58] Gharehbaghi, A., A.A. Sepehri and A. Babic. An edge computing method for extracting pathological information from phonocardiogram. In *ICIMTH*, pp. 364–367, 2019.

[59] Gharehbaghi, A., A.A. Sepehri and A. Babic. Forth heart sound detection using backward time-growing neural network. In *International Conference on Medical and Biological Engineering*, pp. 341–345. Springer, 2019.

[60] Gharehbaghi, A., A.A. Sepehri and A. Babic. Distinguishing septal heart defects from the valvular regurgitation using intelligent phonocardiography. *Studies in Health Technology and Informatics*, 270: 178–182, 2020.

[61] Gharehbaghi, A., A.A. Sepehri, M. Lindén and A. Babic. *Intelligent Phonocardiography for Screening Ventricular Septal Defect Using Time Growing Neural Network*, volume 238, pp. 108–111. IOS Press IOS Press IOS Press IOS Press, 2017.

[62] Gharehbaghi, A., A.A. Sepehri, M. Lindén and A. Babic. *A Hybrid Machine Learning Method for Detecting Cardiac Ejection Murmurs*, pp. 787–790. Springer Singapore, Singapore, 2018.

[63] Gharehbaghi, A., A.A. Sepehri, A. Kocharian and M. Lindén. An intelligent method for discrimination between aortic and pulmonary stenosis using phonocardiogram. In *World Congress on Medical Physics and Biomedical*

Engineering, June 7–12, 2015, Toronto, Canada, volume 51 of IFMBE Proceedings, pp. 1010–1013. Springer International Publishing, 2015.

[64] Graves, A., N. Jaitly and A.-R. Mohamed. Hybrid speech recognition with deep bidirectional lstm. In *2013 IEEE Workshop on Automatic Speech Recognition and Understanding*, pp. 273–278. IEEE, 2013.

[65] Gudmundsson, S., T.P. Runarsson and S. Sigurdsson. Support vector machines and dynamic time warping for time series. In *Neural Networks, 2008. IJCNN 2008. (IEEE World Congress on Computational Intelligence). IEEE International Joint Conference on*, pp. 2772–2776, June 2008.

[66] Gupta, L., D.L. Molfese, R. Tammana and P.G. Simos. Nonlinear alignment and averaging for estimating the evoked potential. *IEEE Transactions on Biomedical Engineering*, 43(4): 348–356, 1996.

[67] He, K., X. Zhang, S. Ren and J. Sun. Deep residual learning for image recognition. In *Proceedings of the IEEE Conference on Computer Vision and Pattern Recognition*, pp. 770–778, 2016.

[68] Hedges, L.V. and I. Olkin. *Statistical Methods for Meta-Analysis*. Academic Press, San Diego, 1985.

[69] Hertz, J.A., A. Krogh and R.G. Palmer. *Introduction to the Theory of Neural Computation*. CRC Press, Boca Raton, USA, 1991.

[70] Hettiarachchi, R., U. Haputhanthri, K. Herath, H. Kariyawasam, S. Munasinghe, K. Wickramasinghe, D. Samarasinghe, A. De Silva and C.U.S. Edussooriya. A novel transfer learning-based approach for screening pre-existing heart diseases using synchronized ECG signals and heart sounds. In *2021 IEEE International Symposium on Circuits and Systems (ISCAS)*, pp. 1–5. IEEE, 2021.

[71] Hochreiter, S. and J. Schmidhuber. Long short-term memory. *Neural Computation*, 9(8): 1735–1780, 1997.

[72] Hollander, M. and D.A. Wolfe. *Nonparametric Statistical Methods*. John Wiley & Sons Inc., New York, 1999.

[73] Huerta, R., S. Vembu, K. Muezzinoglu, Mehmet and A. Vergara. Dynamical SVM for time series classification. In *Pattern Recognition*, volume 7476 of *Lecture Notes in Computer Science*, pp. 216–225. Springer Berlin Heidelberg, 2012.

[74] Jaeger, H. Adaptive nonlinear system identification with echo state networks. *Advances in Neural Information Processing Systems*, 15: 609–616, 2002.

[75] Jaeger, H. Echo state network. *Scholarpedia*, 2(9): 2330, 2007.

[76] Jain, A.K., R.P.W. Duin and J. Mao. Statistical pattern recognition: A review. *IEEE Trans. Pattern Analyse and Machine Intel.*, 22: 4–37, 2000.

[77] Jia, Y., E. Shelhamer, J. Donahue, S. Karayev, J. Long, R. Girshick, S. Guadarrama and T. Darrell. Caffe: Convolutional architecture for fast feature embedding. In *Proceedings of the 22nd ACM International Conference on Multimedia*, pp. 675–678, 2014.

[78] Jozefowicz, R., W. Zaremba and I. Sutskever. An empirical exploration of recurrent network architectures. In *International Conference on Machine Learning*, pp. 2342–2350. PMLR, 2015.

[79] Juang, B.-H. and L.R. Rabiner. A probabilistic distance measure for hidden markov models. *ATT Technical Journal*, 64(2): 391–408, 1985.

[80] Juang, B.-H. and L.R. Rabiner. The segmental k-means algorithm for estimating parameters of hidden markov models. *IEEE Transactions on Acoustics, Speech, and Signal Processing*, 38(9): 1639–1641, 1990.

[81] Kecman, V. *Learning and Soft Computing: Support Vector Machine, Neural Networks and Fuzzy Logic Models*. MIT Press, Cambridge, MA, 2002.

[82] Kim, M. and V. Pavlovic. Sequence classification via large margin hidden markov models. *Data Min. Knowl. Disc.*, 23: 322–344, 2011.

[83] Kohonen, T. Self-organized formation of topologically correct feature maps. *Biological Cybernetics*, 43(1): 59–69, 1982.

[84] Krizhevsky, A., I. Sutskever and G.E. Hinton. Imagenet classification with deep convolutional neural networks. Advances in Neural Information Processing Systems, 25, 2012.

[85] Kubat, M. Neural networks: A comprehensive foundation by simon haykin, macmillan, 1994, isbn 0-02-352781-7. *The Knowledge Engineering Review*, 13(4): 409–412, 1999.

[86] Latif, S., M. Usman and J.Q.R. Rana. Abnormal heartbeat detection using recurrent neural networks. *arXiv preprint arXiv:1801.08322*, 2018.

[87] LeCun, Y., L. Bottou, Y. Bengio and P. Haffner. Deep learning. *Nature*, 521(7553): 436–444, 2015.

[88] LeCun, Y., L. Bottou, Y. Bengio and P. Haffner. Gradient-based learning applied to document recognition. *Proceedings of the IEEE*, 86(11): 2278–2324, 1998.

[89] LeCun, Y. Generalization and network design strategies. *Connectionism in Perspective*, 19(143-155): 18, 1989.

[90] Lee, C-Y., P. W. Gallagher and Z. Tu. Generalizing pooling functions in convolutional neural networks: Mixed, gated, and tree. In *Artificial Intelligence and Statistics*, pp. 464–472. PMLR, 2016.

[91] Lee, K.F. and H.W. Hon. Speaker-independent phone recognition using hidden markov models. *IEEE Trans. Acoustics, Speech and Signal Processing*, 37: 298–305, 1989.

[92] Liang, M. and X. Hu. Recurrent convolutional neural network for object recognition. In *Proceedings of the IEEE Conference on Computer Vision and Pattern Recognition*, pp. 3367–3375, 2015.

[93] Liang, M., X. Hu and B. Zhang. Convolutional neural networks with intra-layer recurrent connections for scene labeling. Advances in Neural Information Processing Systems, 28, 2015.

[94] Lin, T., B.G. Horne, P. Tino and C.L. Giles. Learning long-term dependencies in narx recurrent neural networks. *IEEE Transactions on Neural Networks*, 7(6): 1329–1338, 1996.

[95] Maass, W., T. Natschläger and H. Markram. Real-time computing without stable states: A new framework for neural computation based on perturbations. *Neural Computation*, 14(11): 2531–2560, 2002.

[96] Mallat, S. *A Wavelet Tour of Signal Processing, Third Edition: The Sparse Way*. Academic Press, Inc., USA, 3rd edition, 2008.

[97] McCulloch, W.S. and W.H. Pitts. A logical calculus of the ideas immanent in nervous activity. *Bulletin of Mathematical Biophysics*, 5: 115–133, 1943.

[98] Meintjes, A., A. Lowe and M. Legget. Fundamental heart sound classification using the continuous wavelet transform and convolutional neural networks.

In *2018 40th Annual International Conference of the IEEE Engineering in Medicine and Biology Society (EMBC)*, pp. 409–412. IEEE, 2018.

[99] Messner, E., M. Zöhrer and F. Pernkopf. Heart sound segmentation—An event detection approach using deep recurrent neural networks. *IEEE Transactions on Biomedical Engineering*, 65(9): 1964–1974, 2018.

[100] Michalek, S., M. Wagner and J. Timmer. A new approximate likelihood estimator for arma-filtered hidden markov models. *IEEE Transactions on Signal Processing*, 48(6): 1537–1547, 2000.

[101] Mozer, M.C. Induction of multiscale temporal structure. In *Advances in Neural Information Processing Systems*, pp. 275–282, 1992.

[102] Nagy, G. Feature extraction on binary patterns. *IEEE Transactions on Systems Science and Cybernetics*, 5(4): 273–278, 1969.

[103] Nakagawa, S. and H. Nakanishi. Speaker-independent english consonant and japanese word recognition by a stochastic dynamic time warping method. *IETE Journal of Research*, 34(1): 87–95, 1988.

[104] Oppenheim, A.V., R. Shafer and J.R. Buck. *Discrete-Time Signal Processing*. Prentice Hall, Upper Saddle River, New Jersey, USA, 1998.

[105] Oppenheim, A.V., A.S. Willskey and N.S. Hamid. *Signals and Systems*. Pearson New International Edition, USA, 2014.

[106] Oskiper, T. and R. Watrous. Detection of the first heart sound using a time-delay neural network. In *Proc. Comput. Cardiol.*, volume 29, pp. 537–540, 2002.

[107] Pascanu, R., T. Mikolov and Y. Bengio. On the difficulty of training recurrent neural networks. In *International Conference on Machine Learning*, pp. 1310–1318. PMLR, 2013.

[108] Rabiner, L. A tutorial on hidden markov models and selected applications in speech recognition. *Proceedings of the IEEE*, 77(2): 257–286, 1989.

[109] Rabiner, L. and S. Levinson. A speaker-independent, syntax-directed, connected word recognition system based on hidden markov models and level building. *IEEE Transactions on Acoustics, Speech, and Signal Processing*, 33(3): 561–573, 1985.

[110] Rabiner, L., S. Levinson, A. Rosenberg and J. Wilpon. Speaker independent recognition of isolated words using clustering techniques. *IEEE Transactions on Acoustics, Speech, and Signal Processing*, 27(4): 336–349, 1979.

[111] Rabiner, L., A. Rosenberg and S. Levinson. Considerations in dynamic time warping algorithms for discrete word recognition. *IEEE Transactions on Acoustics, Speech, and Signal Processing*, 26(6): 575–582, 1978.

[112] Rabiner, L. and B.-H. Juang. *Fundamentals of Speech Recognition*. Prentice Hall, United States ed edition, 1993.

[113] Rabiner, L.R. A tutorial on hidden markov models and selected applications in speech recognition. *Proceedings of the IEEE*, 77(2): 257–286, 1989.

[114] Renna, F., J. Oliveira and M.T. Coimbra. Deep convolutional neural networks for heart sound segmentation. *IEEE Journal of Biomedical and Health Informatics*, 23(6): 2435–2445, 2019.

[115] Richard, M. and R. Lippmann. Neural network classifiers estimate bayesian a posteriori probabilities. *Neural Computation*, 87: 1738–1752, 1991.

[116] Sakoe, H. and S. Chiba. Dynamic programming algorithm optimization for spoken word recognition. *IEEE Transactions on Acoustics, Speech, and Signal Processing*, 26(1): 43–49, 1978.

[117] Scherer, D., A. Müller and S. Behnke. Evaluation of pooling operations in convolutional architectures for object recognition. In *International Conference on Artificial Neural Networks*, pp. 92–101. Springer, 2010.

[118] Selim, S.Z. and M.A. Ismail. K-means-type algorithms: A generalized convergence theorem and characterization of local optimality. *IEEE Transactions on Pattern Analysis and Machine Intelligence*, PAMI-6(1): 81–87, 1984.

[119] Semmlow, J.L., M. Akay and W. Welkowitz. Noninvasive detection of coronary artery disease using parametric spectral analysis methods. *Engineering in Medicine and Biology Magazine, IEEE*, 9(1): 33–36, 1990.

[120] Sepehri, A.A., A. Kocharian, A. Janani and A. Gharehbaghi. An intelligent phonocardiography for automated screening of pediatric heart diseases. *Journal of Medical Systems*, 40(1), 2015.

[121] Sepehri, A.A., A. Gharehbaghi, T. Dutoit, A. Kocharian and A. Kiani. A novel method for pediatric heart sound segmentation without using the ECG. *Computer Methods and Programs in Biomedicine*, 99(1): 43–48, 2010.

[122] Sepehri, A.A., J. Hancq, T. Dutoit, A. Gharehbaghi, A. Kocharian and A. Kiani. Computerized screening of children congenital heart diseases. *Computer Methods and Programs in Biomedicine*, 92(2): 186–192, 2008.

[123] Simard, P.Y., D. Steinkraus, J.C. Platt. Best practices for convolutional neural networks applied to visual document analysis. In *Icdar*, volume 3, 2003.

[124] Simonyan, K. and A. Zisserman. Very deep convolutional networks for large-scale image recognition. *arXiv preprint arXiv:1409.1556*, 2014.

[125] Sinha, R.K., Y. Aggarwal and B.N. Das. Backpropagation artificial neural network classifier to detect changes in heart sound due to mitral valve regurgitation. *J. Med. Sys.*, 31: 205–209, 2007.

[126] Sloin, A. and D. Burshtein. Support vector machine training for improved hidden markov modeling. *IEEE Trans. Signal Proc.*, 56: 172–188, 2008.

[127] Stoer, J. and R. Bulirsch. *Introduction to Numerical Analysis*. Springer, 2002.

[128] Sugiyama, M., H. Sawai and A.H. Waibel. Review of tdnn (time delay neural network) architectures for speech recognition. In *1991 IEEE International Symposium on Circuits and Systems (ISCAS)*, pp. 582–585 vol. 1, 1991.

[129] Sujadevi, V.G., K.P. Soman, R. Vinayakumar and A.U. Prem Sankar. Deep models for phonocardiography (pcg) classification. In *2017 International Conference on Intelligent Communication and Computational Techniques (ICCT)*, pp. 211–216. IEEE, 2017.

[130] Sutskever, I., O. Vinyals and Q.V. Le. Sequence to sequence learning with neural networks. In *Advances in Neural Information Processing Systems*, pp. 3104–3112, 2014.

[131] Szegedy, C., W. Liu, Y. Jia, P. Sermanet, S. Reed, D. Anguelov, D. Erhan, V. Vanhoucke and A. Rabinovich. Going deeper with convolutions. In *Proceedings of the IEEE Conference on Computer Vision and Pattern Recognition*, pp. 1–9, 2015.

[132] Sörnmo, L. and P. Laguna. *Bioelectrical Signal Processing in Cardiac and Neurological Applications*. Elsevier Academic Press, USA, 2005.

[133] Tahir, M.A., A. Bouridane and F. Kurugollu. Simultaneous feature selection and feature weighting using hybrid tabu search/k-nearest neighbor classifier. *Pattern Recognition Letters*, 28: 438–446, 2007.

[134] Takami, J.I. and S. Sagayama. A pairwise discriminant approach to robust phoneme recognition by time-delay neural networks. In *[Proceedings] ICASSP 91: 1991 International Conference on Acoustics, Speech, and Signal Processing*, pp. 89–92 vol. 1, 1991.

[135] Tang, H., T. Li and T. Qiu. Noise and disturbance reduction for heart sounds in cycle-frequency domain based on nonlinear time scaling. *IEEE Trans. Biomed. Eng.*, 27: 234–243, 2010.

[136] Tang, Y., X. Wu and W. Bu. Deeply-supervised recurrent convolutional neural network for saliency detection. In *Proceedings of the 24th ACM international conference on Multimedia*, pp. 397–401, 2016.

[137] Taniguchi, T., H. Yamakawa, T. Nagai, K. Doya, M. Sakagami, M. Suzuki, T. Nakamura and A. Taniguchi. A whole brain probabilistic generative model: Toward realizing cognitive architectures for developmental robots. *Neural Networks*, 150: 293–312, 2022.

[138] Tappert, C. and S. Das. Memory and time improvements in a dynamic programming algorithm for matching speech patterns. *IEEE Transactions on Acoustics, Speech, and Signal Processing*, 26(6): 583–586, 1978.

[139] Trentin, E. and M. Gori. Robust combination of neural networks and hidden markov models for speech recognition. *IEEE Transactions on Neural Networks*, 14(6): 1519–1531, 2003.

[140] Vapnik, V.N. An overview of statistical learning theory. *IEEE Trans. Neural Net.*, 10: 988–999, 1999.

[141] Vapnik, V.N. *The Nature of Statistical Learning Theory*. Springer, New York, 2000.

[142] Vembu, S., A. Vergara, M.K. Muezzinoglu and R. Huerta. On time series features and kernels for machine olfaction. *Sensors and Actuators B: Chemical*, 174: 535–546, 2012.

[143] Waibel, A., T. Hanazawa, G. Hinton, K. Shikano and K.J. Lang. Phoneme recognition using time-delay neural networks. *IEEE Transactions on Acoustics, Speech, and Signal Processing*, 37(3): 328–339, 1989.

[144] Wang, B., Y. Lei, T. Yan, N. Li and L. Guo. Recurrent convolutional neural network: A new framework for remaining useful life prediction of machinery. *Neurocomputing*, 379: 117–129, 2020.

[145] Webb, A.R. *Statistical Pattern Recognition*. Hodder Arnold Publication, 1999.

[146] Williams, R.J. and D. Zipser. A learning algorithm for continually running fully recurrent neural networks. *Neural Computation*, 1(2): 270–280, 1989.

[147] Wilpon, J. and L. Rabiner. A modified k-means clustering algorithm for use in isolated work recognition. *IEEE Transactions on Acoustics, Speech, and Signal Processing*, 33(3): 587–594, 1985.

[148] Wilpon, J.G., L.R. Rabiner, C.-H. Lee and E.R. Goldman. Automatic recognition of keywords in unconstrained speech using hidden markov models. *IEEE Transactions on Acoustics, Speech, and Signal Processing*, 38(11): 1870–1878, 1990.

[149] Wissel, T., T. Pfeiffer, R. Frysch, R.T. Knight, E.F. Chang, H. Hinrichs, J.W. Rieger and G. Rose. Hidden markov model and support vector machine based decoding of finger movements using electrocorticography. *Journal of Neural Engineering*, 10: 1–19, 2013.

[150] Wood, J.C. and D.T. Barry. Time-frequency analysis of the first heart sound. *IEEE Eng. Med. Biol. Mag.*, 95: 144–151, 1995.

[151] Wu, X., P. Rozycki and B.M. Wilamowski. A hybrid constructive algorithm for single-layer feed forward networks learning. *Neural Networks and Learning Systems, IEEE Transactions on*, 26(8): 1659–1668, 2015.

[152] Xiao, B., Y. Xu, X. Bi, W. Li, Z. Ma, J. Zhang and X. Ma. Follow the sound of children's heart: A deep-learning based computer-aided pediatric chds diagnosis system. *IEEE Internet of Things Journal*, 7(3): 1994–2004, 2019.

[153] Yamakawa, H. The whole brain architecture approach: Accelerating the development of artificial general intelligence by referring to the brain. *Neural Networks*, 144: 478–495, 2021.

[154] Yamashita, R., M. Nishio, R.K.G. Do and K. Togashi. Convolutional neural networks: an overview and application in radiology. *Insights into Imaging*, 9(4): 611–629, 2018.

[155] Yan, H., Y. Jiang, J. Zheng, C. Peng and Q. Li. A multilayer perceptron-based medical decision support system for heart disease diagnosis. *Expert Systems with Applications*, 30(2): 272–281, 2006.

[156] Yildiz, I.B., H. Jaeger and S.J. Kiebel. Re-visiting the echo state property. *Neural Networks*, 35: 1–9, 2012.

[157] Zavaliagkos, G., Y. Zhao, R. Schwartz and J. Makhoul. A hybrid segmental neural net/hidden markov model system for continuous speech recognition. *IEEE Transactions on Speech and Audio Processing*, 2(1): 151–160, 1994.

[158] Zhang, X. Y. Zou, S. Li and S. Xu. A weighted auto regressive lstm based approach for chemical processes modeling. *Neurocomputing*, 367: 64–74, 2019.

[159] Zhang, X., L.-G. Durand, L. Senhadji, H.C. Lee and J.-L. Coatrieux. Time-frequency scaling transformation of the phonocardiogram based of the matching pursuit method. *Biomedical Engineering, IEEE Transactions on*, 45(8): 972–979, 1998.

Index

For Product Safety Concerns and Information please contact our EU
representative GPSR@taylorandfrancis.com
Taylor & Francis Verlag GmbH, Kaufingerstraße 24, 80331 München, Germany

www.ingramcontent.com/pod-product-compliance
Lightning Source LLC
Chambersburg PA
CBHW070715220326
41598CB00024BA/3169

9 7 8 1 0 3 2 4 1 8 8 6 5